SpringerBriefs on Pioneers in Science and Practice

Volume 28

Series editor

Dr. Hans Günter Brauch, Mosbach, Germany

For further volumes:
http://www.springer.com/series/10970
http://www.afes-press-books.de/html/SpringerBriefs_PSP.htm

Ernst Ulrich von Weizsäcker
Editor

Ernst Ulrich von Weizsäcker

A Pioneer on Environmental, Climate and Energy Policies

Editor
Ernst Ulrich von Weizsäcker
University of Freiburg
Emmendingen
Germany

The cover photograph was taken by James Badham, Santa Barbara, who granted permission for its use in this volume.

Copyediting: PD Dr. Hans Günter Brauch, AFES-PRESS e.V., Mosbach, Germany

ISSN 2194-3125   ISSN 2194-3133   (electronic)
ISBN 978-3-319-03661-8   ISBN 978-3-319-03662-5   (eBook)
DOI 10.1007/978-3-319-03662-5
Springer Cham Heidelberg New York Dordrecht London

Library of Congress Control Number: 2014937721

© The Author(s) 2014
This work is subject to copyright. All rights are reserved by the Publisher, whether the whole or part of the material is concerned, specifically the rights of translation, reprinting, reuse of illustrations, recitation, broadcasting, reproduction on microfilms or in any other physical way, and transmission or information storage and retrieval, electronic adaptation, computer software, or by similar or dissimilar methodology now known or hereafter developed. Exempted from this legal reservation are brief excerpts in connection with reviews or scholarly analysis or material supplied specifically for the purpose of being entered and executed on a computer system, for exclusive use by the purchaser of the work. Duplication of this publication or parts thereof is permitted only under the provisions of the Copyright Law of the Publisher's location, in its current version, and permission for use must always be obtained from Springer. Permissions for use may be obtained through RightsLink at the Copyright Clearance Center. Violations are liable to prosecution under the respective Copyright Law.
The use of general descriptive names, registered names, trademarks, service marks, etc. in this publication does not imply, even in the absence of a specific statement, that such names are exempt from the relevant protective laws and regulations and therefore free for general use.
While the advice and information in this book are believed to be true and accurate at the date of publication, neither the authors nor the editors nor the publisher can accept any legal responsibility for any errors or omissions that may be made. The publisher makes no warranty, express or implied, with respect to the material contained herein.

Printed on acid-free paper

Springer is part of Springer Science+Business Media (www.springer.com)

Foreword

Ernst Ulrich von Weizsäcker: Motor and Path Breaker for Earth Politics

New ideas and innovations are developed by people with an independent mind and the courage to explore new paths. Ernst Ulrich von Weizsäcker is characterized by such independence—together with his positive energy in approaching and communicating with people, and a powerful language that turns complex matters into vivid images.

His personality has made him an exceptional wanderer between the worlds of science, politics, and society—a sustainability researcher, a university president, a national and international environmental politician, who over the past 40 years has had a lasting conceptual and institutional impact on the national and global environmental debate.

Ernst Ulrich is a true member of the "von Weizsäcker" family: his father Carl Friedrich was a philosopher and physicist, his grandfather after whom Ernst has been named was a diplomat (whose role during the Nazi regime has been a contentious issue), his uncle was President of Germany, his brother Carl Christian is a leading economist and headed the German Monopolies Commission. The courage to "think big" has been part of his family heritage from early on. Anyone who as a student in his teens is used to playing table tennis in his parents' house with Heisenberg and other leading intellectuals of the 1950s and 1960s is probably inclined to develop an open and strong mind.

It can nearly be taken for granted that after finishing school, his studies and doctoral research led him directly to his first professorship for biology at the University of Essen in 1972 at the age of 33.

With his appointment as the president of the newly founded University of Kassel in 1975, Ernst began to prepare the ground for a new political agenda that takes the ecological boundaries of the planet seriously and aims at developing perspectives for a desirable future. During his 5 years of presidency, Kassel became one of the few pioneer universities for interdisciplinary environmental research, which it still is today.

He moved on to become director at the UN Centre for Science and Technology for Development in New York. Ernst further developed an international perspective and became director of the Institute for European Environmental Policy in Bonn, London, and Paris.

During this time he wrote his book *Earth Politics*. It was published in 1989 and became a compass for many of us, committed to sustainability, in this period of change at the end of the 1980s and early 1990s. The distinctive feature of this book was the way it provided a new perspective on environmental debates that was future-oriented and global—with convincing concepts and written in a powerful language.

In 1991, the *Wuppertal Institute for Climate, Environment, Energy* was founded. It was brought to life by the visions of two great men: Johannes Rau, at that time Minister-President of North Rhine-Westphalia (and later President of Germany from 1999 to 2004), wanted to create a leading think-tank for sustainability issues. He chose his home town Wuppertal as the location for this institute. Ernst was appointed as the first president of the institute and was granted the freedom to build an institute that can make a difference in the sustainability debate.

Ernst lived up to that challenge. In Wuppertal, he created an environment for innovative and controversial thinking. He deliberately appointed other lateral thinkers and established radical interdisciplinarity as the normal mode of operation at the Wuppertal Institute. This led to the development of new visions of resource productivity, sustainable transport policy, or an ecological tax reform, which influenced the debates of the 1990s. Ernst has always been aware of the fact that for science to have an impact on society, vivid images and vigorous language are needed, in order to convey its important messages. Until this day, "Factor 10" by Friedrich Schmidt-Bleek, the "Factor 4" by Ernst Ulrich von Weizsäcker or the study "Sustainable Germany" coordinated in 1996 by Reinhard Loske and Raimund Bleischwitz have proven to be concise and vivid concepts that influence the debate about the potential of increasing resource efficiency and a comprehensive understanding of sustainable development. The fact that the Wuppertal Institute is well received and welcomed on the international stage until this day is owed to this founding period and the influence of Ernst.

In 1998, he found his way into the German Parliament as a member of the Social Democrats—a daring change of sides from science to politics. Until 2005 he was a member of the Parliament and actively pursued the goal of establishing a perspective on "earth politics" in the German political sphere—as Chairman of the Parliament's *Study Commission on Economic Globalization* and from 2002 onwards as Chairman of the Parliament's Environmental Committee. The implementation of the red–green coalition government's ecological tax reform took place during that time. Ernst was directly elected to represent his Stuttgart constituency in 2002 and he was a popular and publicly visible politician. Nonetheless, the tension between his visionary and proactive approach on the one hand, and the day-to-day routines of political business in Berlin on the other were sometimes hard to bear for him.

At the age of 66 he returned to the international arena: in 2005 he was appointed, effective 2006, Dean of the *Bren School for Environmental Science and Management*, UCSB in Santa Barbara, California, USA. About 30 years after the

University of Kassel had been founded, Ernst again created an environment for interdisciplinary sustainability research—this time in a US state with progressive environmental politics. In 2007, Ernst became Co-Chair of UNEP's *International Resource Panel*—a global panel, similar to the IPCC that compiles the state of research and develops policy recommendations for the decoupling of future economic development and consumption of resources. Since 2012—exactly 40 years after the publication of the famous study on the *Limits to Growth* and 40 years after the commencement of his first professorship—Ernst holds one of the two Chairs of the Club of Rome. He dedicates himself to this position with the same spirit and positive energy that have characterized him from early on and led him to be an inspiration for young people committed to sustainability just as much as for the Chinese government in the context of his consultancy mandates.

We are thankful for the innovative concepts Ernst has created, his ability to create spaces for unconventional thinking and his seemingly infinite energy that has had a lasting impact on the international sustainability debate!

January 2014　　　　　　　　　　　　　　　　　　　　　　　　Uwe Schneidewind

Prof. Dr. Uwe Schneidewind (Germany) is President of the Wuppertal Institute for Climate, Environment an Energy since 2010 and Professor for Sustainable Transition Management at the University of Wuppertal. He studied Business Administration at the University of Cologne and HEC/Paris, worked and wrote his Ph.D. and Habilitation on strategic environmental management issues at the University of St. Gallen in Switzerland. Uwe Schneidewind held a full-time Professorship position for Production Management and the Environment at the University of Oldenburg (1998–2010) and was President of the University (2004–2008).

Acknowledgments

I wish to thank Hans Günter Brauch, the editor of this book series, for initiating this volume and for doing all the work of a good editor. I myself would never have considered putting together old scribblings of mine and getting them re-published.

I wish to thank my fabulous co-authors who typically did what I would never have done:

- My wife *Christine*, apart from being a wonderful partner and companion, had the ingenious ideas behind redefining information as novelty and introducing the term error-friendliness (something schoolteachers tend to hate but should love);
- *Irene Kehler*, who drafted the submission to OECD of the first international exposure of my wild ideas in the early 1970s—35 years before "Bologna"—of a modular university system;
- *Jochen Jesinghaus*, who looked at all the relevant statistics based on which we can now realistically talk about long-term price elasticity;
- *M. S. Swaminathan* and *Aklilu Lemma*, who shaped the work of the Advisory Committee of the UN Commission on Science and Technology for Development in such a way that it truly made sense for developing countries;
- *Friedrich "Bio" Schmidt-Bleek*, not actually a co-author but an inspiring partner at the Wuppertal Institute, where he introduced the "factor" concept that induced me to think big in terms of energy and resource productivity;
- *Amory* and *Hunter Lovins*, who were and still are the real gurus of energy efficiency and of the business case for it;
- *Oran Young* and *Matthias Finger*, who understand the economic and philosophical frame behind privatization and what is needed to minimize its obvious downsides;
- *Karlson "Charlie" Hargroves* and his fabulous Australian TNEP team, who took the initiative of updating Factor 4 and in the end reinvented and transformed it into Factor 5.

I could mention dozens of other friends, teachers, and partners who influenced and supported my walks through a long life. I can mention but a few. My father, Carl Friedrich von Weizsäcker (PSP 20, 21, 22, 23, 24, 25), was probably the first to waken scientific curiosity in me. Among the many academic teachers I enjoyed, I would like to mention Bernhard Hassenstein, zoologist at Freiburg University, who took the risk of accepting a physicist for a Ph.D. dissertation in biology. Georg Picht (PSP forthcoming) guided me into interdisciplinary research in his unique Heidelberg institute. Johannes Rau, then Minister of Science in North Rhine–Westphalia (and later president of Germany), lured me into the founding senate of the new University of Essen, and I was somehow appointed as full professor of interdisciplinary biology. Hans Krollmann in the government of Hesse made me the founding president of Kassel University, at the time a bit of a wild institution trying out what would never flourish in classical, discipline-centred universities. It was a tumultuous time for me as well, but I fondly remember great friendships with many professors and students.

From Kassel, I moved to the UN in New York, where I was privileged to work in a division with two outstanding senior staff members, Munirathna Anandakrishnan and Aklilu Lemma, who soon became my friends and indeed guides to the policy challenges that developing countries were and are facing. Without their compassion and wisdom I would never have acquired the knowledge and sensitivity needed for (later) co-chairing the *UNEP International Resource Panel*. One day, I received a call from Europe from my friend Konrad von Moltke, asking me if I would consider taking over from him as director of the *Institute for European Environmental Policy* (IEEP). I said yes, applied, was appointed, and plunged into a new world of experience, greatly helped by IEEP's team which consisted of (among others) Nigel Haigh, David Baldock from London, Alain Sagne, Thierry Lavoux, and later Brice Lalonde from Paris, and Pascale Kromarek, Annie Roncerel and Hubert Meiners in Bonn. The institute served as a great learning space for me to understand the functioning of the EEC (now the EU) and the role of parliaments in shaping international environmental policies. An Advisory Board had as its members the famous Edgar Faure, and Adriaan Oele, Lord Cranbrook, Raymond Georis, and Wolfgang Burhenne, and later also Klaus Töpfer and Beate Weber. A better advisory board could not have been thought of.

A good 15 years after my tenure as biology professor, Johannes Rau, in the meantime Prime Minister of North Rhine–Westphalia, appointed me as founding president of the Wuppertal Institute, doubtless the most enjoyable and indeed successful position in my professional life, and a paradise, I should say, for lateral thinking. The Wuppertal team, likewise, was unique and catapulted the institute to top visibility in Germany and worldwide. A few names may be chosen to stand for this success: Bio Schmidt-Bleek, as already mentioned, Peter Hennicke, the great energy efficiency guru, who took over the presidency when I left, Wolfgang Sachs, Reinhard Loske and Manfred Linz, who masterminded the highly influential book on "Sustainable Germany," Rudolf Petersen, who chaired and inspired the highly unorthodox transport division, Jill Jäger, leading the climate division, Harry Lehmann who brought solid systems analysis into the house, and Hans Kretschmer and Wolfram Huncke who taught us to visualize and publicize our academic findings—making the institute a favourite partner for the media, for schools, and for policymakers. And then there was that fabulous group of "second" level scholars, most of whom later made it to the first level. Among them were Manfred Fischedick, Raimund Bleischwitz, Christa Liedtke, Stefan Bringezu, Fritz

Hinterberger, Karl-Otto Schallaböck, Meike Spitzner, Kora Kristof, Stephan Thomas, Stephan Lechtenböhmer, and Jochen Luhmann, the highly spirited historian of modern environmental policy, who together with Kai Schlegelmilch made the institute an important voice in ecological tax reform. Finally, there was Maryse Biermann, who saw to it that the chaos of lateral thinking at the institute was always balanced by the reliable functioning of the president's office. Also the Wuppertal Institute was greatly helped by an outstanding International Advisory Board, chaired by Prof. Hartmut Graßl, Hamburg, an eminent climatologist, and including, among others, Prof. Robert Ayres (Resources for the Future), Jim MacNeill, Ottawa (former Secretary General of the Brundtland Commission), Freda Meissner-Blau, Vienna (then President, Ecoropa), Florentin Krause, Berkeley, USA, and Pier Vellinga and Wouter van Dieren, both from Amsterdam.

Later, in the Bundestag, I went through another stage of learning, this time from political leaders who included (from the SPD) Gerhard Schröder, Michael Müller, Peter Struck, and Ulrike Mehl. In the Enquete Commission on Globalization, I had a great team that included Jochen Boekhoff, Marianne Beisheim, and Achim Brunnengräber who did the background work for a successful political dispute on and research into the challenges of globalization (leading eventually to the writing of *Limits to Privatization*).

At the Bren School in Santa Barbara, my local work, including fundraising and curriculum development, did not lead to any major written output. But it so happened that during that time, I ran into Charlie Hargroves in Beijing, and we decided to write an update of *Factor 4* that later developed into *Factor 5*. It would have been impossible for me to keep on working on the book had I not been helped by Greg Bratman, a student and later graduate of the Bren School, who became enthused with working on this leading-edge manuscript, and also helped me keep pace with the Australian team—who wrote twice as many pages as we could possibly include in the book. (We put the excess pages into a web page accessible to readers of the book).

Back in Germany in 2009, I was singularly lucky in moving with Christine into an extremely comfortable, energy-neutral new house, together with our daughter Paula Bleckmann, her husband Frank, and their three sons Benedikt, Elias, and Anselm. The house is large enough to accommodate our entire extended family at Christmas or other family reunion times: our son Jakob and his wife Eva with their children Ulrich, Joseph, Paul, and Franziska, our son Adam with his wife Tanja and daughters Mia and Anna, our son Franz with his son Luca, and our daughter Maria, unmarried so far. I am immensely grateful to Christine and the entire family for allowing so much fun and entertainment including very serious discussions about daily issues and the challenges of our century.

The perspective of this book is indeed the challenges of the century, and I dedicate the book to the generation of our grandchildren!

Emmendingen, Germany, March 2014 Ernst Ulrich von Weizsäcker

Contents

Part I Ernst Ulrich von Weizsäcker

1 **Autobiographical Sketch**................................. 3

2 **Ernst Ulrich von Weizsäcker's Selected Publications**.......... 7

Part II Key Texts by Ernst Ulrich von Weizsäcker

3 **Modular University**..................................... 21
   Ernst Ulrich von Weizsäcker and Irene Kehler
   3.1 Introduction...................................... 21
   3.2 The Principles Guiding the Curricula of a BGHS......... 22
   3.3 Curricular Implications and Problems of an Integrated
        Short-Cycle Higher Education System.................. 25
        3.3.1 Equality of Opportunity for Access............. 25
        3.3.2 Courses of Studies.......................... 26
        3.3.3 Duration of Studies......................... 28
        3.3.4 Examinations............................... 28
        3.3.5 Compactness of Course....................... 29
        3.3.6 Small Groups and Tutoring................... 29
        3.3.7 Learning by Discovery....................... 30
        3.3.8 Compulsory Knowledge....................... 31
        3.3.9 New Information and Communication
               Techniques................................ 31
        3.3.10 Permanent Education........................ 32
   References... 33

4 **Contagious Knowledge: Contagion as a Quality Criterion
   for Disciplinary and Interdisciplinary Science**................ 35
   4.1 Introduction...................................... 35
   4.2 Scientific Success................................. 36
   4.3 Contagious Information............................. 37

	4.4 Contagious Science	42
	4.5 Quality and Responsibility	47
	References	51

5 New Frontiers in Technology Application: Integration of Emerging and Traditional Technologies ... 53
  5.1 Preface ... 53

6 The Environmental Dimension of Biotechnology ... 57
  6.1 Eurosclerosis and Beyond ... 57
  6.2 Biotechnology: Environmental Pros ... 58
    6.2.1 Agriculture ... 58
    6.2.2 Industry ... 59
  6.3 Environmental Cons: Success is the Danger ... 61
    6.3.1 Evolutionary Concerns ... 62
    6.3.2 Some Thoughts on Regulation ... 62
  References ... 65

7 Not a Miracle Solution but Steps Towards an Ecological Reform of the Common Agricultural Policy (CAP) ... 67
  7.1 Introduction ... 67
  7.2 Sixteen Elements of an Ecological Reform of the CAP ... 69
    7.2.1 Food Quality and Consumer Information ... 69
    7.2.2 Limiting the Use of Certain Chemicals ... 70
    7.2.3 Limiting Intensive Livestock ... 71
    7.2.4 Nitrates ... 71
    7.2.5 Applying and Enforcing Existing Environmental Legislation ... 72
    7.2.6 Compensation Payments ... 73
    7.2.7 Extensification Schemes ... 74
    7.2.8 Payments for Ecological Work ... 75
    7.2.9 Set-Aside ... 75
    7.2.10 Prepension Scheme ... 77
    7.2.11 Biotechnology ... 77
    7.2.12 Part-Time Agriculture ... 78
    7.2.13 Hobby Agriculture ... 80
    7.2.14 Organic Farming ... 80
    7.2.15 Direct Marketing ... 81
    7.2.16 Food Security Policy and Active Price Policy ... 81
    7.2.17 Preliminary Assessment of the Package ... 82
  7.3 Financial Implications ... 82
    7.3.1 Political Acceptance ... 83
  References ... 84

Contents

8 Earth Politics 87
    8.1   Introduction: Why Earth Politics? 87
            8.1.1   The Economic Century: An Episode 89
            8.1.2   The Rape of Nature 91
            8.1.3   Earth Politics for the Century
                      of the Environment 93
    8.2   Ecological Taxes and Charges 95
    References 97

9 Ecological Tax Reform 99
Ernst Ulrich von Weizsäcker and Jochen Jesinghaus
    9.1   The Size of the Challenge 99
    9.2   Ecological Tax Reform 101
    9.3   Price Elasticity 102
            9.3.1   First Stage of Adjustment 102
            9.3.2   Second Stage of Adjustment 103
            9.3.3   Third Stage of Adjustment 103
            9.3.4   Fourth Stage of Adjustment 103
            9.3.5   Fifth Stage of Adjustment 104
    9.4   Price Elasticity for Fuels: A New Measurement Concept 105
            9.4.1   Long Term Price Elasticity of Fuel
                      Consumption is High 105
    9.5   Objections and Obstacles to Ecological Tax Reform 108
            9.5.1   The Fiscal Policy Objection: The Conflicting
                      Goals of Yield Versus Steering Effect 108
            9.5.2   The Social Policy Objection: 'Green Taxes
                      Aren't Fair' 109
            9.5.3   The Environmental Policy Objection: 'Revenue
                      Should Be Purpose-Linked' 110
            9.5.4   The Economic Policy Objection: 'Ecological
                      Taxes Area Burden on the Economy' 111
            9.5.5   The Voter Objection: 'Ecological Taxes
                      Aren't Popular' 112
            9.5.6   The Effectiveness Objection: 'There are Instruments
                      Which Work Faster and with Greater Precision
                      Than Ecological Taxes' 113
            9.5.7   The Harmonization Objection: 'Ecological Taxes
                      Cannot Be Harmonized Within the EC' 114
            9.5.8   Objections from the Losers 115
    9.6   What Makes Ecological Tax Reform Attractive
            for the Business World? 115
    References 117

10	**Sustainable Energy Policies: Political Engineering of a Long Lasting Consensus**		119
	10.1	An Alliance of Long-Termers	119
	10.2	Greenhouse Effect	120
	10.3	A New Direction for Technological Progress	121
	10.4	Ecological Tax Reform	122
	10.5	Transport Costs and International Trade	124
	References		125
11	**Factor Four: Doubling Wealth—Halving Resource Use: A New Report to the Club of Rome**		127
	Ernst Ulrich von Weizsäcker, Amory B. Lovins and L. Hunter Lovins		
	11.1	Preface	127
	11.2	Introduction: More for Less	128
		11.2.1 Exciting Prospects for Progress	128
		11.2.2 Moral and Material Reasons	129
	11.3	Chapter 1: Twenty Examples of Revolutionising Energy Productivity	129
	11.4	Chapter 2: Twenty Examples of Revolutionising Material Productivity	131
	11.5	Chapter 3: Ten Examples of Revolutionising Transport Productivity	132
		11.5.1 Videoconferences	132
	11.6	Chapter 11: We May Have 50 Years Left to Close the Gaps	134
		11.6.1 Beyond the Limits? The Meadows May be Right	135
		11.6.2 Population Dynamics	138
		11.6.3 What has 'Factor Four' Got to Do with the Population?	139
	References		141
12	**Eco-Efficiency Goals: Factor Four or Factor Ten**		143
	12.1	Why Should Factor Four and Factor Ten Co-exist?	143
	12.2	Which Are the Differences Among the Two?	144
	12.3	Which Are the Appropriate Time Horizons?	145
	12.4	What Should be Done Next?	145
	12.5	Which Can be the Role of Governments?	146
	References		147
13	**Sharing the Planet: From 'Limits to Growth' to 'Factor Four'**		149
	13.1	From 'Limits to Growth' to Sustainable Development	149
	13.2	Biodiversity and Dematerialization	151
	13.3	Climate Change and the Energy Dilemma	152

	13.4	After the Industrial Revolution the Eco-Efficiency Revolution	153
	13.5	The Good News: Factor Four	154
	13.6	Profitability, Long Term and Short Term	156
	References		157
14	Globalization and Its Challenges to Democracy and Development		159
15	Limits to Privatization		167

Ernst Ulrich von Weizsäcker, Oran Young and Matthias Finger

	15.1	Introduction: Seeking a Balance	167
	15.2	A Word on Definitions	168
	15.3	Changing Role of the State	169
	15.4	Forms of Privatization	170
	15.5	Points of Departure: Polar Perspectives	172
	15.6	'Horses for Courses'	174
	15.7	Generic Pros and Cons	174
	15.8	The Shape of Things to Come	176
	References		177
16	Information, Evolution, and 'Error-Friendliness'		179

Ernst Ulrich von Weizsäcker and Christine von Weizsäcker

	16.1	Shannon's Information	179
	16.2	Novelty and Confirmation	180
	16.3	Genetic Information	183
	16.4	Evolution	185
	16.5	Error-Friendliness	187
	References		190
17	Factor Five: Transforming the Global Economy Through 80 % Improvements in Resource Productivity		193

Ernst Ulrich von Weizsäcker, Karlson 'Charlie' Hargroves, Michael H. Smith, Cheryl Desha and Peter Stasinopoulos

	17.1	Introduction: Factor 5—The Global Imperative	193
		17.1.1 Balancing Economic Aspirations with Ecological Imperatives	195
	17.2	A Long-Term Ecological Tax Reform	202
		17.2.1 Introduction	202
		17.2.2 Two Centuries of Falling Resource Prices	204
		17.2.3 The Concept of a Long-Term Ecological Tax Reform	208
		17.2.4 Overcoming the Dilemma of Short-Term Instruments	209

		17.2.5	The Poor, the Blue-Collar Workers, the Investors, and the Fiscal Conservatives	210
	References			213

18 Climate Sceptics Keep a Distance from Solutions ... 215
References ... 217

19 Decoupling: Technological Opportunities and Policy Options: Executive Summary ... 219
    19.1   Introduction ... 219
    19.2   Trends in Resource Use ... 220
    19.3   Consequences of These Changes ... 221
    19.4   Strategic Implications ... 222
    19.5   Choices of Response for Policymakers ... 223
    19.6   Technological Responses Allowing Significant Decoupling ... 226
    19.7   Creating the Conditions for Investments in Resource Productivity ... 228
    19.8   Changing Current Biases ... 228
    19.9   'Lock-In' to Political and Economic Structures ... 230
    19.10   Making Progress with Resource Productivity ... 231
    19.11   Unlocking Change in Policies ... 231
    19.12   Changing the Institutional Framework to Facilitate Future Policy Reform ... 233
    19.13   Putting Decoupling into Practice: Linking Resource Price Rises to Resource Productivity Gains ... 234
    19.14   Broadening the Economic Discourse ... 234
    19.15   Creating a Vision of the Future and Reducing Uncertainty ... 235
    19.16   Creating Sufficient Winners in Favour of Change ... 236
    19.17   Taking Account of Potential Losers in a Policy Mix ... 236
    19.18   Creating New Institutional Arrangements ... 237
    References ... 237

UNEP's International Resource Panel ... 239

Club of Rome ... 241

Federation of German Scientists ... 243

Wuppertal Institute on Environment, Climate and Energy ... 245

University of Freiburg ... 247

About the Coauthors 251

About the Author 253

About the Book 255

Part I
Ernst Ulrich von Weizsäcker

Ernst Ulrich von Weizsäcker speaking as a member of the German Parliament in the Bundestag. Reprint is permitted for purposes of political education. © Deutscher Bundestag, MELDEPRESS/ Benjamin Hofgarten

Chapter 1
Autobiographical Sketch

I was born on 25 June, 1939 in Zürich, Switzerland and grew up in Göttingen, Germany where I graduated from the Max-Planck-Gymnasium. Later, I studied chemistry and physics at the University of Hamburg (Physics Diploma 1965). I switched to Biology and moved to Freiburg University, (Ph.D.: 1969). Since 1969, I am married with Christine von Weizsäcker, née Radtke, also a biologist and policy advisor. We have five children and so far ten grandchildren.

1965–1968   Graduate student of zoology and biological cybernetics at the University of Freiburg with Prof. Bernhard Hassenstein. Doctoral thesis on the neuro-cybernetics of the form vision of honey bees.

1968–1972   Member of the Board of the SPD Baden-Württemberg, with a special focus on university reform. Proposal for a new curricular concept of higher education using modules and encouraging interdisciplinary interaction among students. Publication: *Baukasten gegen Systemzwänge. Der Weizsäcker Hochschulplan* (See Chap. 3).

1969–1972   Senior Fellow, Protestant Research Institute, Heidelberg. Director: Prof. Georg Picht. Peace research (on biological weapons disarmament and control and on the ecological dimension of peace); and theoretical work on information theory and open systems. Main publications: *BC-Waffen und Friedenspolitik*, and *Humanökologie und Umweltschutz*. Besides, engagement on the reform of higher education.

1972–1975   Full Professor for Interdisciplinary Biology, University of Essen, and member of the university's Founding Senate. Aside from the full time work on the build-up of the new university teaching duties in animal physiology, evolution, ecology, and mathematical biology. At the same time editing of the first serious scientific book, called *Offene Systeme I*, on the temporal structures of information, entropy, and evolution.

1975–1980   Founding President, University of Kassel. Study reform engagement. Creation of several interdisciplinary research centres including the first one in Germany on organic farming. No scientific publications under my name.

1981–1984	Director for Policy, Analysis and Research at the UN Centre for Science and Technology for Development. Servicing the Advisory Committee on Science and Technology for Development; creation of the Advance Technology Alert System (ATAS) with a regular ATAS journal. Creation of an international working group on the integration of very modern and traditional technologies. Publication, with M.S. Swaminathan and Aklilu Lemma as co-editors: *New Frontiers in Technology Application. On the integrated application of emerging and traditional technologies.*
1984–1991	Director, Institute for European Environmental Policy, Bonn, London, Paris, a non-governmental think tank with some 8 professional staff in 1984. Core funding (sufficient for 5 staff) came from a Dutch foundation. I expanded the institute to some 30 staff and established new offices in Brussels and in Arnhem, Netherlands. We pioneered the study of implementation at national levels of European environmental legislation and we were the first in the policy analysis of East European environmental problems. We advised governments, parliaments, the European Commission and private business on international environmental policy questions, including the greening of the Common Agricultural Policy. During one full academic year, from 1987 to 1988, I also served as guest professor at the Technological University of Darmstadt, teaching on environmental policy, and on the theory of open systems. The policy lectures were later printed as a book, *Erdpolitik* that was translated into many foreign languages including English (*Earth Politics*). In its core, the book sketched out the transition from pollution control to a broader understanding of environmental policy that would include energy, transport, agriculture, biotechnology, climate, resource flows, and North–South relations on environmental matters.
1991–2000	Founding President, Wuppertal Institute for Climate, Environment and Energy. Generously supported by the North-Rhine Westphalian government, this think tank soon became an international leader in radical approaches to eco-efficiency (*Factor Four, Factor Ten*). *Factor Ten* was the radical concept of the Institute's Vice President, Prof. Friedrich Schmidt-Bleek. His resonance was best in Japan and later in China, but also international bodies including OECD, UNDP, and the EU-Commission often referred to these concepts. In 2004, I wrote a book, *Factor Four,* on this radical approach, together with Dr. Amory B. Lovins and his then wife L. Hunter Lovins. The book was written in English but first published in German and in that translation was on the bestseller lists for half a year.
1998–2005	Member of the Bundestag (German Parliament), SPD, for Stuttgart. From 1999 to 2002 I served as Chairman of the *Bundestag Study Commission on Economic Globalization.* This Committee had a high

visibility in the country and partly abroad. When the ILO created the *World Commission on the Social Dimensions of Globalisation*, I was appointed a member and became actively involved. In 2002, I was re-elected to the Bundestag by conspicuously winning the local majority for my party in the local Stuttgart constituency. Afterwards, I was elected as Chairman of the *Bundestag Environment Committee*. Duties included agenda setting, pushing for Europeanization of German environmental policies, establishing international contacts. I decided as early as 2002 not to run for a third term. From 2001 to 2004, I had some time to steer and co-ordinate the writing of a book on *Limits to Privatization* that was published in 2005. Co-editors were Prof. Oran Young from Santa Barbara and Prof. Matthias Finger of the European Business School in Lausanne, Switzerland. It was accepted as a report to the Club of Rome.

2006 Dean of the Bren School for Environmental Science and Management, UCSB, Santa Barbara, California, USA. Here my duties included fundraising and the development towards a broader agenda beyond nature conservation and related economics. In particular, I introduced the notion of resource productivity and helped create a professors chair on the subject. I was also appointed professor at the School, teaching primarily on resource productivity. From the outset, I said I was accepting the Dean's functions only for 3 years.

Ernst Ulrich von Weizsäcker (*second from left*) as a student of physics, with his siblings in the Austrian Alps, 1962. From *left* Carl Christian, later a famed professor of economics; Ernst Ulrich; Heinrich, later a professor of mathematics, and Elisabeth, historian, later married to Konrad Raiser, and a leader in protestant churches lay movement

In 2007, Achim Steiner, Executive Director of UNEP, appointed me as Co-Chair of the newly created *International Panel for Sustainable Resource Management*, later renamed *International Resource Panel*. I am still serving in this function, probably until 2016. The Panel has established itself as a major international authority on metals including their recycling, on land use, on environmental damages caused by resource extraction and use, and generally on the decoupling of economic well-being from resource consumption.

Another assignment, from 2007 to 2009 was that of co-chairing a Task Force of the *China Council for International Cooperation on Environment and Development* (CCICED). The Task Force, composed of 50 % Chinese and 50 % international experts, focused on economic instruments for energy efficiency and the environment. Our Report, published in 2009, advocated a slow but steady price increase of energy and mineral resources,—in line with the documented increase of resource productivity,—so that on average monthly bills for energy and materials would not increase.

Also in 2007, when attending a conference in Beijing, I met with Mr. Charlie Hargroves from Australia, who said that *Factor Four* was one of his favourites,— asking if it was not time to think of an update. In effect, after consulting Amory and Hunter Lovins, I agreed with Charlie Hargroves and his Australian team to write a new book, now called *Factor Five*. Not a single line was left from Factor Four, but the spirit is the same. Like Factor Four, also Factor Five was accepted as a Report to the Club of Rome. Many translations were made and published of both books. Factor Five was certainly more mature in its policy segments. It advertised the idea of the slow but steady increase of energy and mineral prices.

In January, 2009, I returned to Germany, to live with my family in a newly built energy neutral house in Emmendingen, near Freiburg. I received more and more requests for speaking engagements, mostly on Factor Five and resource productivity.

In April, 2011, the University of Freiburg appointed me as Honorary Professor, with mild, non-remunerated teaching duties.

In October, 2012, I was elected Co-President of the *Club of Rome*, together with Dr. Anders Wijkman, former member of the European Parliament from Sweden. When we took over, the Club of Rome was nearly bankrupt. We knew it but heard inklings of a major Chinese donation to the Club, with some strings attached. It materialized and provided a perspective of a medium term financial stability. In 2014, we initiated the revival of a "young Club of Rome" group.

Awards 1978: *Pfaff Prize for Education*; 1989: *Premio de Natura* (Italy), shared with PM Gro Harlem Brundtland; 1991: Honorary Professor, Technological University of Valparaiso, Chile; 1991: Member, Club of Rome; 1996: *Duke of Edinburgh Gold Medal of WWF International*; 2000: Honorary Degree, Soka University, Japan; 2001: *Takeda Award*, shared with. F. Schmidt-Bleek; 2008: *German Environment Prize*; 2009: *Grand Cross of Merit of Germany* (Großes Bundesverdienstkreuz); 2010, Honorary Degree, Belgrade University; 2011: *Theodor Heuss Prize*; 2012: *Medal of Merit*, Baden Württemberg. In 2013, the Gottlieb Duttweiler Institute listed me among the 100 Top Thought Leaders (measuring impacts in English language communities).

Homepage www.ernst.weizsaecker.de and English website on this book: <http://www.afes-press-books.de/html/SpringerBriefs_PSP.htm>.

Chapter 2
Ernst Ulrich von Weizsäcker's Selected Publications

1966

"Untersuchungen zu zwei apparativen Verbesserungen in der Gammastrahlen-Diagnostik". Physik-Diplom (M.Sc.) dissertation with the chair of biophysics (Prof. H.A. Künkel), Medical department, University of Hamburg, Germany.

1970

"Dressurversuche zum Formensehen der Bienen, insbesondere unter wechselnden Helligkeitsbedingungen", in: *Z. vergl. Physiologie* (renamed as *Journal of Comparative Physiology*), 69: 296–310.

(Editor and co-author): B*C-Waffen und Friedenspolitik* (Stuttgart: Klett).

1971

(Editor, with others, and lead author): *Baukasten gegen Systemzwänge. Der Weizsäcker-Hochschulplan* [*Academic Building Blocks Against Curricular Straight Jackets. The Weizsäcker University Plan*] (Munich: Piper).

1972

"Wiederaufnahme der begrifflichen Frage. Was ist Information?", in: *Nova Acta Leopoldina*, 37,1: 535–555.

(Editor and co-author): *Humanökologie und Umweltschutz* (Stuttgart: Klett).

1974

(Editor and co-author): *Offene Systeme I. Beiträge zur Zeitstruktur von Information, Entropie und Evolution* (Stuttgart: Klett); 2nd ed., 1986.

1974–1984

Between 1974 and 1984 publications were mostly anonymous due to my functions as university president (in Kassel, Germany) and as a UN civil servant, respectively.

1984

(Co-editor with Swaminathan, M.S.; Lemma, Aklilu, and co-author): *New Frontiers in Technology Application. Integration of Emerging and Traditional Technologies* (Dublin: Tycooly International).

(with Weizsäcker, Christine von): "Fehlerfreundlichkeit", in: Kornwachs, Klaus (Ed.): *Offenheit, Zeitlichkeit, Information: Zur Theorie der Offenen Systeme* (Frankfurt–New York: Campus): 167–201.
"Überraschung und Bestätigung gibt Information", in: Schaefer, Gerhard (Ed.): *Information und Ordnung* (Cologne: Aulis): 89–98.

1985
(Editor and co-author): *New Technologies—Clean Industry?* (Bonn: Institute for European Environmental Policy).
"Sind neue Technologien umweltfreundlich? Chancen und Gefahren in Landwirtschaft und Industrie", in: *Ökologische Konzepte*, 21: 9–16.

1986
"Contagious knowledge: Contagion as a quality criterion for disciplinary and interdisciplinary science", in: Ganelius, Tord (Ed.): *Progess in Science and Its Social Conditions*. Nobel Symposium 58 (Oxford–New York: Pergamon): 171–182.
"Qualitatives Wachstum. Eine Skizze zur Auseinandersetzung mit Ilya Prigogine/Isabelle Stengers 'Dialog mit der Natur'", in: Altner, Günter (Ed.): *Die Welt als offenes System* (Frankfurt: Fischer): 48–54.
(Editor and co-author): *Waschen und Gewässerschutz. Ein Konflikt kommt zur Sprache.* (Karlsruhe: Müller).
"Die Zukunft wird anders als wir denken", in: Hafemann, Michael; Schlüpen, Detlef (Eds.): *Technotopia* (Weinheim–Basel: Beltz): 9–18.
"Change in the philosophy of work", in: *Int. J. Production Research*, 24: 743–748.
(with Weizsäcker, Christine von): "Biologie und Technik. Fehlerfreundlichkeit als Evolutionsprinzip und Kriterium der Technikbewertung", in: *Universitas*, 41: 791–800.
"Umweltschutz als europäische Aufgabe", in: *Lutherische Monatshefte*: 264–267.
"The environmental dimension of biotechnology", in: Davies, Duncan (Ed.): *Industrial Biotechnology in Europe* (Brussels: Centre for European Policy Studies—European Commission): 35–45.

1987
(Co-author with Weizsäcker, Christine von): "How to Live With Errors. On the Evolutionary Power of Errors", in: *World Futures. The Journal of General Evolution*, 23,3: 225–235.
"Ganzheitlicher Umweltschutz—eine Herausforderung für Politik und Informatik", in: Jaeschke, A.; Page, B. (Eds.): *Informatikanwendungen im Umweltbereich* (Berlin–Heidelberg–New York: . Springer): 1–7.
"Elemente einer europäischen Einigung in der Agrar-Umweltpolitik", in: Urff, W. von (Ed.): *Landwirtschaft, Umwelt und Ländlicher Raum* (Baden-Baden: Nomos): 273–296.
Not a Miracle Solution but Steps Towards an Ecological Reform of the Common Agricultural Policy (Bonn: Institute for European Environmental Policy).

"Umweltschutz. Eine neue Dimension der internationalen Politik", in: Genscher, Dietrich (Ed.): *Nach vorn gedacht. Perspektiven deutscher Außenpolitik* (Stuttgart: Bonn Aktuell): 195–210.

1988

(Editor and co-author): *Gutes Trinkwasser, wie schützen? Konflikt um Wasserversorgung und Gewässerschutz* (Karlsruhe: Müller).
"Public Environmental Consciousness—How Does It Affect European Industry?" CON-CAWE Senior Management Seminar of 8–9 October, 1987 (The Hague: CONCAWE): 21–28.
"Die Gefahren der Gentechnologie liegen in ihrem flächendeckenden Erfolg" [Commentary on the deliberate release of genetically modified plants], in: *Universitas*, 43: 1146–1149.
"Fehlerfreundlichkeit lebender Systeme", in: Universität des Saarlandes (Ed.): *Saarbrücker Universitätsreden 29*: 21–26.
(with Schreiber, Helmut): "Luftreinhaltung-Der schwierige Konsens", in: Gündling, Lothar; Weber, Beate (Eds.): *Dicke Luft in Europa* (Karlsruhe: Müller): 163–172.

1989

"Die Umweltkrise—eine Herausforderung für unser Bildungssystem", in: BMBW (Ed.): *Umweltbildung in der EG* (Bad Honnef: Bock): 28–38.
"Internationale Harmonisierung im Umweltschutz durch ökonomische Instrumente", in: *Jahrbuch zur Staats-und Verwaltungswissenschaft*, 3: 203–216.
"Binnenmarkt und Umwelt", in: *Scheidewege*, 19: 91–98.
(Co-author with Simonis, Udo Ernst): "Global environmental problems", in: *The Crisis of Global Environment Demands for Global Politics* (Bonn: Foundation for Development and Peace): 13–35.
"Economics and the environment. New strategies for the European countryside", in: *Land Use Policy*, 6: 295–300.
Erdpolitik. Ökologische Realpolitik an der Schwelle zum Jahrhundert der Umwelt. (Darmstadt: Wissenschaftliche Buchgesellschaft). Five updated editions were published in German until 1997. This book was translated into five languages including English (1994).

1990

"Die Preise müssen die Wahrheit sagen", in: *Universitas*, 45: 133–143 (reprinted in English as: "Prices should tell the truth", in: *Universitas*, 1992: 297–304).
"Über die Umwelt Bescheid wissen: Wettbewerbsfaktor für Unternehmen und Staaten", in: *VDI-Berichte*, 825: 99–114.
"Die GATT-Uruguay-Runde ökologisch beeinflussen", in: *Agrarische Rundschau*, Vienna.
Entwicklung der Umweltpolitik in EG und Osteuropa—Schritte zu einer ökologischen Marktwirtschaft (St. Gallen: ÖBU).

"Global warming and environmental taxes", in: *Int. J. of Global Energy Iss*ues, 2: 14–19.

"Wer zahlt den Preis der Umweltzerstörung?", in: Erdölvereinigung (Ed): *Wirtschaft, Wachstum, Risiko* (Zurich): 35–53.

Leaders of the Wuppertal Institute for Climate, Environment and Energy. From left to right Professors Manfred Fischedick, Vicepresident since 2006, Uwe Schneidewind, President since 2010, Ernst von Weizsäcker, founding President from 1991 to 2000, and Peter Hennicke, from 2010 to 2008. With permission of the Wuppertal Institute granted in March 2014

1991

(Editor and co-author): *Weniger Abfall—gute Entsorgung. Konflikte um den Hausmüll* (Karlsruhe: Müller).

"Sustainability: A Task for the North", in: *Journal of International Affairs*, 44,2: 421–432.

1992

"Ökologische, soziale und gesellschaftliche Forderungen an die Landwirtschaft", in: *Landwirtschaft in der Industriegesellschaft* (Frankfurt a. M.: DLG-Verlag).

"Von der Umwelttechnik zur umweltverträglichen Technologie: eine neue technologische Revolution steht bevor", in: *GAIA*: 272–277.

"Towards an Alliance Between Environmentalists and Insurers", in: *Environment and Insurance* (Cologne: Gerling): 14–20.

(Co-author with Jesinghaus, Jochen): *Ecological Tax Reform. A Policy Proposal for Sustainable Development* (London–Atlantic Highlands, NJ: Zed Books).

(Co-editor and co-author, with Bleischwitz, Raimund): *Klima und Strukturwandel* (Bonn: Economica Verlag).

"Vorstellung des Wuppertal Instituts", in: *Klima und Strukturwandel*: 21–37.
"L' économie du marché est-elle vraiment plus propre?", in: *Geopolitique*, 40: 43–47.
"Von der Umwelttechnik zur umweltverträglichen Technologie. Eine neue technologische Revolution steht bevor", in: *GAIA*: 272–277.
"Ökologischer Strukturwandel als Antwort auf den Treibhauseffekt", in: *Aus Politik und Zeitgeschichte*, B 16: 33–38.

1993

(Co-author with Petersen, Rudolf): "Mobility in the greenhouse", in: *UNEP Industry and Environment*, 16,1–2: 7–10.
"Wachstumsgrenzen und neue Wachstumschancen", in: *Nova Acta Leopoldina*, NF 69, No 285: 331–347.
"Zum Demokratieproblem", in: Göhner, Reinhard (Ed.): *Die Gesellschaft für morgen* (Munich: Piper): 139–151.
"Ecological structural change as a response to the greenhouse effect", in: *Economics*, 48: 34–42.
"Arbeitsproduktivität versus Energieproduktivität: Perspektiven und Grenzen der Steigerung von Produktivität als Motor der Wachstumsgesellschaft", in: *Jahrbuch Arbeit und Technik*. (Bonn: Dietz Verlag).

1994

Earth Politics (London: Zed) (translated and updated from the German *Erdpolitik*, 1989).
1994–2011: (Co-editor): *Jahrbuch Ökologie* (Yearbook on national and international environmental policies, Munich: C.H. Beck [until 2007]; Stuttgart: Hirzel Verlag [since 2008]) as a regular contributing author.
"Kreislauf als Weg zum Aufschwung", in: *Schweizerische Technische Zeitung* (STZ), 5/94: 27–32.
"Wohlstand im Jahrhundert der Umwelt", in: *Universitas*, 1: 1–11.
"How to Achieve Progress Towards Sustainability?", in: *Sustainable Consumption* (Oslo: Ministry of Environment).
"Umwelterziehung war erst der Anfang. Die Umweltkrise verlangt uns mehr ab", in: *DGU Nachrichten* (Hamburg), 10,94.
"Sustainable Economy", in: *The Science of the Total Environment* (Elsevier), 143, 149–156. SSDI 0048-9697(94)003703-5.
"Ökologische Effizienzrevolution: eine Chance für Mitteleuropa" [text translated into Czech] (Prague: Environment Centre, Charles University Prague): 5–21.

1995

"Negawatts, Lifestyles and Incentive Structures". Invited Paper to the World Energy Council 16th Congress, Tokyo, Session 3.1: 11–25.
(Co-author with Lovins, Amory; Lovins, Hunter). *Faktor Vier. Doppelter Wohlstand, halbierter Naturverbrauch* (Munich: Droemer Knaur) [The book was a Report to the Club of Rome and was translated into twelve languages (English edition. See 1997).

1996

"Effizienzrevolution und intelligente Informationstechnik", in: *Der GMD-Spiegel*, 12/96: 42–45.

"Ethik der Globalisierung", in: *Ethik-Letter—Lay Report* (Frankfurt), L4: 1–3.

(Co-author with Luhmann, Jochen): "Joint Implementation", in: *Energiewirtschaftliche Tagesfragen* (Essen): 354–355.

1997

(Co-author with Lovins, Amory; Lovins, Hunter): *Factor Four. Doubling Wealth, Halving Resource Use* (London, Earthscan).

"Die Spannung zwischen Globalisierung und Umweltschutz", in: *Heidelberger Club* (Ed.) *Globalisierung* (Berlin: Springer): 15–23.

(Editor and co-author): *Grenzen-Los? Jedes System braucht Grenzen—aber wie durchlässig müssen diese sein?* (Basel–Boston–Berlin: Birkhäuser).

Desafíos del medio ambiente y respuestas políticas (Santiago de Chile, Estudios Públicos, Primavera): 149–169.

"Wer fällt in die Globalisierungsfalle?", in: *Scheidewege*, 27: 348–352.

1998

"Wirtschaftspolitisches Forum: Ökologische Steuerreform", in: *Zeitschrift für Wirtschaftspolitik* (Stuttgart: Lucius & Lucius): 47: 70–77.

(Co-author with Weizsäcker, Christine von): "Information, evolution and 'error-friendliness'", in: *Biological Cybernetics*: 501–506.

"Generationenvertrag in Gefahr!", in: *LayReport*: 1–2.

"Ein technologischer Ruck für das 21. Jahrhundert", in: Lafontaine, Oskar; Schröder, Gerhard (Eds.): *Innovationen für Deutschland* (Göttingen: Steidl): 52–58.

1999

(Co-author with Quarch, Christoph): *Eine neue Politik für die Erde. Die globale Partnerschaft von Wirtschaft und Ökologie* (Freiburg: Herder).

"Wider den Sozialdarwinismus", in: *Ökologisch-evolutionäre Reflexionen. Neue Sammlung.* (Seelze: Friedrich Verlag): 531–542.

(Editor and co-author): *Das Jahrhundert der Umwelt* (Frankfurt–New York: Campus).

(Co-author with Bleischwitz, Raimund): "Green Productivity", in: *New Economy* (London: Blackwell), 6: 40–43.

"Nachhaltigkeit. Neue Technologien, neue Zivilisation", in: Grossmann, W.D.; Eisenberg, W.; Meiss, K.M.; Multhaup, T. (Eds.): *Nachhaltigkeit. Bilanz und Ausblick* (Frankfurt: Peter Lang): 63–74.

"Die Energieeffizienz vervierfachen—erneuerbare Energien vervielfachen", in: *Energie Extra*, 4/99: 1.

"Die Umweltzerstörung ist nicht etwa verlangsamt, sondern beschleunigt worden", in: Jelich, F.J.; Schneider, G. (Eds.): *Politik und Politische Bildung für das 21. Jahrhundert* (Düsseldorf: DGB).

"Ökologische Jahrtausendwende in Japan", in: *Zeitschrift für Angewandte Umweltforschung* (ZAU), 12: 455–457.

Ernst Ulrich von Weizsäcker during a hearing of the Enquete Commission of the German Parliament on Growth, Welfare and Life Quality in the year 2011. To the right are Prof. Reinhard Hüttl and Ms Daniela Kolbe MdB, chairperson of the Enquete Commission. Reprint is permitted for purposes of political education. © Deutscher Bundestag, Achim Melde

2000

"Erwerbsarbeit in der Dritten Lebensphase", in: Dahlmanns, G. (Ed): *Prosperität in einer alternden Gesellschaft* (Frankfurt: Frankfurter Institut): 187–202.

2001

(Co-editor, with Stigson, Björn; Seiler-Hausmann, Jan-Dirk): *From Eco-Efficiency to Overall Sustainable Development* (Wuppertal: Wuppertal Institute).

2002

"Mitwelt, Nachhaltigkeit und Erdpolitik", in: Ingensiep H.W., Eusterschulte, Anne (Eds.): *Philosophie der natürlichen Mitwelt. Festschrift für Klaus Michael Meyer-Abich* (Würzburg: Könishausen and Neumann): 241–245.
"Globalisierung und Demokratie", in: *Scheidewege*: 238–244.

2003

"Gedanken über den Nutzen von Grenzen", in: Hogrebe, Wolfram (Ed.): *Grenzen und Grenzüberschreitungen* (Berlin: Akademie-Verlag): 451–456.
"Wir müssen die Demokratie neu erfinden", in: *Natur & Kosmos* (June): 38–42.

2004

(Co-editor, with Seiler-Hausmann, J.-D.; Liedtke, Christa): *Eco-efficiency and Beyond. Towards the sustainable enterprise* (Sheffield: Greenleaf Publ.).
"Visionen zu DaimlerChrysler 2020", in: *AK Umwelt der MitarbeiterInnen der DaimlerChrysler AG* (Stuttgart: DaimlerChrysler AG): 102–103.

"The Old and New Europe: Alternatives for Future Transatlantic Relations?", in: Bohne, Eberhard; Bonser, Charles F.; Spencer, Kenneth M: (Eds.): *Transatlantic Perspectives on Liberalization and Democratic Governance* (Hamburg: Lit): 384–396.

(Co-author with Stürmer, Martin): "Grenzen der Privatisierung", in: *Scheidewege*, 34: 29–42.

2005

"Produktivität, angewandt auf Chemie und Energie. Editorial", in: *Chemie in unserer Zeit*, (Weinheim: Wiley VCh): 303.

(Co-author and lead editor, with Young, Oran; Finger, Matthias): *Limits to Privatization. How to avoid too much of a good thing* (London: Earthscan).

"Freie Fahrt für die Wirtschaft, -Schluss mit dem Umweltschutz? Die Effizienzrevolution versöhnt Ökonomie und Ökologie", in: Altner, G. et al. (Eds.): *Jahrbuch Ökologie 2006*: 154–158.

2006

Grenzen der Privatisierung. Wann ist des Guten zuviel? (update and translation of Limits to Privatization; Stuttgart: Hirzel Verlag). The book was also translated into Chinese.

2007

"Complexity, Interdisciplinarity and Overview: Virtues of 21st Century Universities", in: Festschrift at the occasion of the 80th birthday of Professor Mircea Maliţa. Published in *Millennium III*, a Review of the Black Sea University Foundation, Bucharest, Romania, ISSN 1454-7759.

"New Opportunities for the Bren School". Kick-off presentation for a Strategy Meeting on the Bren School's role on the UCSB campus. Campus circulation only.

2008

"Economic Instruments for Energy efficiency and the Environment". (Co-author). Interim Report of the China Council for International Cooperation on Environment and Development (CCICED), Beijing.

"New Direction for Technological Progress" Science Dialogue on Energy, Santa Barbara, June 2008, later circulated by IHDP (International Human Dimensions Programme on Global Environmental Change), Bonn.

(With Peter Goldmark) "The Decarbonization Challenge—US and European Perspectives on Climate Change". Transatlantic Thinkers Series. Bertelsmann Foundation, Gütersloh.

2009

(With Hargroves, Charlie, et al.): *Factor Five. Transforming the Global Economy through 80 % Improvements in Resource Productivity* (London: Earthscan).

2010

Faktor 5. Die Formel für nachhaltiges Wachstum [updated, thoroughly edited and translated into German of "Factor Five"] (Munich: Droemer-Knaur). [The Chinese edition, published in 2010 (Shanghai: Wisdom Press) is based on the German update; so are the Russian edition, 2012, the French edition, 2013, and the Japanese edition, 2014].

"Energy Efficiency", in: Juergensmeyer, Mark (Ed.): *Encyclopedia of Global Studies* (London: Sage).

"Was schrumpft bei 'Faktor Fünf'? Eine Effizienzstrategie", in: *Politische Ökologie*, 121 (Munich): 48–50.

"Factor Five", in: *UNIDO Proceedings of 2009 Manila Conference on Greening Industry* (Vienna: UNIDO).

"Hoffnungsschimmer Zivilgesellschaft, aber nicht von Amerika gegängelt", in: Brunnen-gräber, Achim (Ed.): *Energieangebotspolitik war gestern*, in: *Technikfolgenab-schätzung. Theorie und Praxis*, 3/2010 (Berlin: Deutscher Bundestag).

"Confronting the Rebound Effect", in: Lehmann, Harry (Ed.): *Factor X: Policy, Strategies and Instruments for a Sustainable resource use* (Dessau: Umweltbundesamt).

2010–2014

Ernst Ulrich von Weizsäcker as SPD Bundestag spokesman for Sustainable Development, at the occasion in 2004 of creation of the Bundestag Sub-Committee on Sustainable Development, with Bundestag President Wolfgang Thierse at the centre. Reprint is permitted for purposes of political education. © Deutscher Bundestag, Achim Melde

Six Dialogues with Daisaku Ikeda, *Institute for Oriental Philosophy, Tokyo 2010–2014*.

2011

"Normen und Anreize", in: Djordje Pinter (Ed.): *Wirtschaft—Gesellschaft—Natur-Ansätze zu einem zukunftsfähigen Wirtschaften-Festschrift für Prof. Dr. Eberhard K. Seifert* (Marburg; Metropolis, 2011): 585–590.

"Kompetenzfeld Natur. Contribution to a volume commemorating the 40th Anniversary of the University of Kassel." in: D. Postlep (Ed.): *Vierzig Jahre Universität Kassel*. (Kassel: University Kassel).

"The Significance for China of the Factor Five Concept", (article was written in English but published in Chinese), in: Manhong Mannie Liu, David Ness, Haifeng Huang (Eds): *Green Economy and Its Implementation in China. A Collection of Essays on Ecological Issues in China* (Shanghai, Wisdom Press).

"How Europe should tackle its resource constraints", in: *Europe's World* (Summer 2011): 4–8.

"Une économie décarbonée est possible", in: *Journal de l'Environnement*, at: http://www.journaldelenvironnement.net/article/une-economie-decarbonee-est-possible, 13/04/2011).

"Technology and Policy Options for Making Transport Systems More Sustainable". With Charlie Hargroves. Background Paper No.2, UN Commission for Sustainable Development CSD19/2011/BP2, New York, 2–13 May 2011.

"Allianz Europa—Asien: Neubeginn der globalen Umweltpolitik", in: Udo E. Simonis et al. (Eds.): *Jahrbuch Ökologie 2012* (Stuttgart: Hirzel).

"Gerechtigkeit, Klima, Umwelt", in: Ottmar Fuchs (Ed.), 2012 *Horizont Weltkirche. Erfahrungen—Themen—Optionen und Perspektiven - Festschrift für Prof. Josef Sayer* (Mainz: Matthias-Grünewald-Verlag).

"How Europe should tackle its resource constraints", in: *Europe's World* (Summer 2011): 4–8.

2012

"Atomstaat und Klimadesaster verhindern", in: Walter Spielmann (Ed.): *"Projekt Zukunft„—Zur Bedeutung Robert Jungks im Wandel der Zeit* (Salzburg: Otto-Müller-Verlag, 2012).

2013

"Boosting resource productivity: Creating ping-pong dynamics between resource productivity and resource prices" (with Robert U. Ayres), in: *Environmental Innovation and Societal Transitions*, 9 (2013): 48–55.

"Ökosoziale Modernisierung", in: Joseph Riegler (Ed.):, *Zukunft als Auftrag* (Mauerbach, Austria: DTW Zukunft).. 121–129.

"Klaus Töpfer 75", in: Falk Schmidt, Nick Nuttal (Ed.): *Klaus Töpfer at 75 Festschrift für Klaus Töpfer* (München; Oekom).

"Das Gebot des Maßhaltens", in Reto Ringger (Ed.) Footprint Yearbook 2013. Zürich: Globalance Bank.

"Effizienz—erforderlich für Nachhaltigkeit" in Udo E. Simonis et al. (Eds.) Mut zu Visionen. Jahrbuch Ökologie 2014. (Stuttgart: Hirzel), 64–70.

"Why not an Efficiency Revolution for China?", in: *Boao Review* (October 2013): 100–101.

2014

(Lead author): "*Decoupling: Technological Opportunities and Policy Options*" (Nairobi: UNEP International Resource Panel).

Ernst Ulrich von Weizsäcker as a member of the German Parliament in the Bundestag. Reprint is permitted for purposes of political education. © Deutscher Bundestag, Foto-und Bildstelle

Part II
Key Texts
by Ernst Ulrich von Weizsäcker

Part II
Key Texts
by Ernst Ulrich von Weizsäcker

Chapter 3
Modular University

Ernst Ulrich von Weizsäcker and Irene Kehler

3.1 Introduction

1. The model of a *Baukasten-Gesamthochschule* (BGHS) represents a complete integration of present short-cycle higher education into university education. This paper there-fore deals with only one model of a linkage between short and long-cycle studies, namely the "integrated comprehensive university model".[1] Yet short and long-cycle studies do not mean quite the same thing in a BGHS as in the definition used in some countries. They do not necessarily imply a vocational or practical orientation, since there may exist both short theoretical courses as well as longer but practically oriented courses of study.
2. The comprehensive model of a BGHS not only facilitates access to post-secondary higher education, it also allows permeability in all directions with a minimum loss of time; simultaneously it implies a reform of methods of learning and teaching. Even though these advantages may appear fascinating at first glance, this should not make us overlook the problems related to its implementation within present structures of higher education, to external and internal factors, to differences of structure of knowledge in various disciplines, and last but not least to the uncertainties resulting from a rather imperfect knowledge of the psychology of learning

---

Note by the Secretariat: The attached paper has been prepared by Mrs. Irene Kehler of the Arbeitsgruppe für empirische Bildungs-forschung, Heidelberg, and slightly revised and amended by Dr. Ernst Ulrich von Weizsäcker, as a contribution to the discussion at a meeting of the OECD (OECD, Directorate for Scientific Affairs meeting on Short-Cycle Higher Education (Grenoble, 15–17 October 1971). This text is in the public domain and is republished within the OECD's publication policies.), with particular reference to the specific item of the agenda concerning the structure and content of studies in short-cycle higher education (OECD, DAS/EID, 71.73, Paris, 22 October 1971).

---

[1] Cf. "Towards New Structures of Post-Secondary Education: A Preliminary Statement of Issues" (Paris: OECD, 1971): 39–40.

3. The BGHS model has now reached the following stage: Since the first proposal for a Baukasten-type university was presented (1969), work was mainly concentrated on theoretical questions. Important features of the system had to be changed, so that the first publication can no longer be considered as a reference. Meanwhile, several German universities became interested in introducing at least parts of the system on an experimental and small-scale basis.

3.2 The Principles Guiding the Curricula of a BGHS

4. The general aims underlying the model of a BGHS (Weizsäcker 1971) should be mentioned before proceeding with the analysis of the principles guiding the curriculum of a BGHS:

- Expansion of the post-secondary sector of higher education at reasonably low cost;
- Equality of opportunity for access to higher education;
- Adoption of new functions such as provision of short-cycle higher education and of permanent education within a multi-functional higher education institution;
- Provision of interdisciplinary curricula for every conceivable combination of fields;
- Education for a world of rapid scientific development and changing occupational demands, i.e. education of graduates capable of independent and continued study and research;
- Education of politically interested graduates capable of independent opinion formation and commitment;
- Provision of both interdisciplinary and conventional research facilities for university teachers and students.

  5. Access to the BGHS will be organized in such a way as to enable students without the formal qualifications (Abitur, baccalaureat, etc.) to enter university. Anybody may take a first course of 4–6 weeks; if successfully completed he may take two more courses which, again, if successfully completed, enable him to take up a regular course of study of 3 years. Eor part-time students these course requirements can partly be met by evening, radio, or television courses. Every student is supposed to consult an academic advisor who recommends adequate courses within the chosen discipline.
  6. The course of studies comes close to an "open curriculum" in the sense that there will be no prescribed courses. In the student's own interest there will be model courses of studies, suggesting but not prescribing certain sequences. At the beginning of their studies, however, students will be required to take a compulsory "orientation course" of 4–6 weeks' duration.
  7. Curriculum decisions will be made by the student in close co-operation with an academic adviser and possibly also with an individual tutor. Students ought to major in one subject, and in the area of this subject they will choose about

one half of their courses. The remaining courses are divided between related and more remote subjects. This organization allows permeability in all directions with a minimum loss of time, because the individual curriculum can easily be adapted to such minimal requirements. The advisory services will enable the student, particularly the inexperienced student, to make use of this permeability and flexibility. These services should provide the best and most up-to-date information on the educational as well as the occupational system. On the other hand, the danger of a misuse of information-power for manipulative purposes might arise, and this should be guarded against.

8. The duration of studies is considered to be 3 years for a short-cycle' course of any type (theoretical or practical). The relatively few students continuing their studies immediately after this period will receive a long-cycle education of, say, 5 years. Because of the rapid changes in knowledge and occupational requirements, all graduates are expected to take up studies every few years (e.g. every 3 years) by completing one course. Students graduating from short-cycle courses taking advantage of this institutionalized permanent education may thus reach the formal qualification related to a long-cycle course of studies.

9. The awarding of a degree will be based on a credit system without final examination. Contrary to present practices credits will have an external value and students can leave the university at any time they wish. Course grades should be differentiated and cover all activities and qualifications so as to constitute a valuable information for potential employers.

10. Compact courses or units of 4–6 weeks' duration (5–6 days a week full-time) or 8–12 weeks duration for half-time students are the elements of a "Baukasten course of study". A year would be divided into 8 periods of about 6 weeks so that each period would contain one "high intensity course". Since a student only takes four courses a year, he can choose among a wide range of courses and has at his disposal some 20–30 weeks of free time for preparation, deepening of understanding and individual or group continuation of course work. The time between courses could also be partly used as a buffer to equalize qualification of rather different participants in a course. This system also makes it easier to take a course in spite of employment, once a paid leave for educational purposes is introduced.

11. Tutoring and studying in small, groups of about 5–7 people will allow active and effective learning, evening courses with several hundred participants. Tutors can be teachers, students, or students returning to university for further education. They will get an adequate salary and possibly also a certain training in group dynamics. Older students, experienced in group learning and team work, will no longer need tutors.

12. Learning by discovery[2] will be the preferred method of study, supplemented by guided discovery or pure information, according to the nature of the

---

[2] This concept was elaborated by the German Bundes-assistentenkonferenz (BAK): *Forschendes Lernen—Wissenschaftliches Prüfen*. Schriften der Bundesassistentenkonferenz 5 (Bonn: Bundesassistentenkonferenz, 1970). Also: L. Huber, in: Weizsäcker, Dohmen, Jüchter, *op. cit.*

discipline and of learning materials. But research and discovery as a way of learning presuppose a new definition of 'research', which was until now an activity reserved for academic professors. There is also need for a new interdisciplinary context which does not necessarily imply new knowledge in the single disciplines involved. Courses will, wherever possible, be focussed on a scientific problem and on problem oriented learning which is suitable for both science and occupation oriented study. By emphasising the solution of (practical or theoretical) problems, it questions the present value system which attributes highest prestige to pure theory and which is the root of today's hierarchy of 'noble' and 'less noble' higher education. The student will, on the contrary, be led to a functional view of the relation between theory and practical application.

13. Compulsory knowledge of the major discipline chosen by the student cannot 'be' aspired to in an open curriculum., Learning by discovery—as a relatively time-consuming way of learning—requires a selection of examples among the abundance of knowledge constituting a discipline. But in order to avoid eclecticism, some six fundamental aspects or basic concepts should be formulated for each discipline, and each of these aspects has to appear at least once during a normal curriculum. This clearly constitutes a restriction in the freedom of choice of study units, but since these 'fundamental aspects' will be treated systematically in a great variety of different units, the restriction is much smaller than in any other curricular system. For example, aspects of biology are: ecology, heredity, molecular processes, metabolism, central nervous functions, and comparative morphology; needless to say, usually more than one aspect would be treated systematically in most units of the BGHS. Transfer to other fields and problems must be learned and exercised, In addition, emphasis is put on knowledge from other disciplines, because an interdisciplinary approach is necessary for the understanding, discussion or solution of most vital problems in industrial society.

14. The use of new information and communication techniques is a prerequisite for the integration of students from distant locations of those in full-time employment as well as for the provision of 'information packages' to be used when they are needed in a course or in individual study. Audiovisual aids and other ways of standardized instruction allow to fill knowledge gaps within short periods of self or group instruction, which is particularly important for students of a lower educational level.

15. An integration of permanent education into the university will replace the actual Reparation between institutions of higher education and institutions of adult education. The integration is conceived globally as in the case of short-cycle or vocationally oriented education. It can therefore affect the composition of courses and small study groups. Permanent education does not need to become a formal requirement; we can hope to motivate adults by various means for a regular return to university (see paragraph 6). Such 'means', incentives, or motivations could, for example, be an excellent provision of courses by radio, TV, and other technical media, courses given in

accommodations near to the university, an appropriate orientation of course subjects, according to the needs of students, a market mechanism sanctioning favourably permanent education, a positive attitude of employers towards permanent education provided by the public education system, the realization of the need for further education, an up-to-date information system and finally, but most important, a paid leave for educational purposes of, say, 3–4 weeks every 3 years.

3.3 Curricular Implications and Problems of an Integrated Short-Cycle Higher Education System

16. In accordance with the terminology used in the introduction, we shall now concentrate on curricular implications of both integration of long- and short-cycle higher education, as well as of vocationally and scientifically oriented studies. Further implications derive from the specific principles of the BGHS, which are discussed in more detail in an intermediate report concerning our project.[3] Our analysis will consider empirical results and experiences wherever possible without going into details of learning psychology. Several discussion issues are followed by conclusions or recommendations. The order of presentation corresponds to Sect. 3.2.

3.3.1 Equality of Opportunity for Access

17. It must be realized that the formal opening of university education to graduates of all kinds of secondary education is not a sufficient condition to achieve this aim. Experience has proved that it is particularly difficult to get blue collar workers and the rural population into university: 66/0 of the first applicants to the British Open University were In non-manual and intellectual occupations[4]; participants in the Bavarian Telekolleg exhibited an overrepresentation of employees and civil servants (by some 150 %) and of participants from large and medium-sized cities (over 10,000 inhabitants), but an underrepresentation of workers (by about 55 %) and of participants from

---

[3] I. Kehler, G. Kellner, E. v. Weizsäcker: Baukasten-studium—Volkswirtschaftslehre und Soziologie. Zwischenbericht des Baukastenprojekts. Diskussions-beiträge der Arbeitsgruppe für empirische Bildungs-forschung, hektographiert (Heidelberg, 1971). In this paper some recent estimation of the costs of a Baukasten university are also provided.

[4] Paper presented to a conference on "Study without Lecture Hall" by Otto Peters in Gummersbach, September 1970.

small towns (less than 1,000 inhabitants) (Bayerischer Rundfunk 1970). This leads to the conclusion that in order to get more potential talent into university, both material and time conditions ought to be improved and curricula better adapted. More difficult, but necessary, is a change in the attitudes towards university education and occupational mobility.

18. An adaptation of curricula seems to be particularly important in all courses qualifying for entry into university, in courses for which there is a relatively low demand (recommended for first-year students) and in courses in which there is a high heterogeneity of participants. 'Adaptation' implies that inductive rather than deductive approaches are used, or that problems raised in a course are related to vital issues of interest to the students. The latter particularly applies to social science courses, political, psychological or vocational courses. Finally, there is another curriculum problem of a correct evaluation and selection of students in these courses. This problem is raised in paragraph 27.

3.3.2 Courses of Studies

19. The rules concerning the sequences within different courses of study represent a key factor determining the degree of permeability or the loss of time for a transfer from one course of studies to another. Although the BGHS model is based on an open curriculum, it appears that prescribed sequences, or at least partially prescribed sequences, are unavoidable in a number of hierarchically-structured 'sciences. Frequently mentioned examples are mathematics and, to some extent, physics, engineering, and perhaps medicine. Today the common characteristic of these subjects is their rather rigid sequence of course requirements as determined by historical factors and by a real or presumed structure of knowledge.

20. In most other disciplines, however, hierarchical structures or substructures play a minor role. Curricula in these fields would be open with only three restrictions:

- First, an introductory orientation course providing both technical information and training, and information on university life and cultural facilities.
- Secondly, in order to guarantee a sufficiently broad education, each "fundamental aspect" (cf. paragraph 13) of a discipline has to be learned at least once if a graduation in this discipline is to be achieved. Because any aspect of a study field could be treated in combination with almost any other aspect in certain-units, a student who wishes to specialize in one aspect could still meet the broad education requirements by choosing those units.
- Thirdly, a student should take only 50 % of the units in his main discipline. Another 25 % should be taken in related fields and methodological sciences such as statistics. The remaining 25 % of the study time should be reserved for units in remote fields, especially in interdisciplinary and in social problems. (For

social scientists the requirements of the latter 25 % would be fulfilled by taking units in technological, medical, or anthropological sciences.)

21. For the more hierarchically-structured sciences, we assume that it would be sufficient to offer the basic concepts and aspects of the respective science in one or several introductory courses. In mathematics these could be: set theory, theory proving, logic, and calculus. Model courses of studies would then offer alternative sequences and recommend certain course prerequisites. All courses, in whatever field, must point out the relations to basic concepts or the relevance of basic concepts for this particular field. Again, in mathematics such fields could be: algebra, topology, probability theory, functional analysis. Different concepts and aspects can thus be touched on in every course and studied with more intensity in general courses on a higher level. This system allows for a spiral and cumulative way of learning.

22. Practically oriented courses of studies such as architecture, medicine, engineering and technical studies, pedagogics and social work, will in addition require work on projects and/or practical work. The well-known problem arises, whether such a course should have a sequence in which theoretical studies precede practical application or vice versa. There are two reasons that favour an inversion of the present preferred sequence from theoretical to practical studies. First this sequence proved highly inefficient, as, for example, in the education of long-cycle engineers. Secondly, in such a case, transfer from short-cycle courses to traditional long-cycle studies is hardly or not at all possible. The solution we want to advocate here for an integrated course seems to be somewhat similar to the Yugoslav model mentioned in the OECD document.[5] (We think that the lack of success of this model can partly be explained by its too radical features of curriculum inversion and, perhaps, by the too high transfer rates due to a lack of labour-market functioning and transparency.)

23. How could an integrated short- and long-cycle course for engineers look? The discussion with an engineer resulted in the following model: the access to university requires the completion of either an apprenticeship for students without higher secondary education, or a year of practical work for the others. Apprenticeship should provide the technical and mathematical knowledge necessary to achieve an equal standard with secondary school graduates; the practical year should provide the necessary practical experiences. The first 3 years of an integrated course of studies consist of a problem oriented scientific education emphasizing projects, but also offering purely theoretical courses. Graduates would be quite comparable to present graduates as to their level of knowledge, but more active, self-reliant and used to team work.

24. The second cycle would offer theoretical and specialized education. As far as the co-ordination of studies within the two cycles is concerned, the remarks made in paragraph 6 apply: there are recommended sequences or prerequisites

---

[5] OECD: "Towards New Structures of Post-Secondary Education", *op. cit.*, para. 137.

as well as frequent references to structures and concepts, and transfer to different problem areas. This example does not claim general validity, the integration of theoretically and practically oriented studies in other fields may demand different models.

3.3.3 Duration of Studies

25. Short-cycle studies cannot follow today's models of short-cycle higher education, the aim of which is to squeeze a maximum amount of learning materials into 2–3 years. Here the German word 'Ausbildung' applies as opposed to 'Bildung' or 'Studium'. In the BGHS model, students of short-cycle higher education should follow the same principles as students of long-cycle higher education, i.e. they should learn to think, to ask questions and to look behind ready-made formulae and rules.
26. Short-cycle courses therefore require the recognition and application of new curricular means accelerating the presentation of information and reducing the amount of learning materials so that enough time is left for learning by discovery. Such curricular methods may be based on audio-visual aids, the reduction of the number of lectures, certain forms of programmed instruction, or even computer-assisted instruction, and. on a selection of learning material. An education that limits itself to the transmission of knowledge to passive students cannot claim to be scientific, nor can it be integrated into university education. Thus short-cycle courses lose their terminal character, since they equip students with the cognitive, motivational and behavioural qualifications needed for a long-cycle course of studies.

3.3.4 Examinations

27. Vocationally oriented, short-cycle courses traditionally, put much weight 011 the acquisition of testable and examinable knowledge. If this is no longer to be the basis of such courses, the function of examinations must change. Their role as a means of selection and provision of information to the outside world is decreased by the inclusion of "learning by discovery" and of new cognitive and non-cognitive aims of studying such as critical and creative thinking, motivation, ability to learn, to communicate and to co-operate, which are not testable or measurable. The main function of examinations should be internal feedback for students and teachers. Having lost much of their former significance, they will 110 longer create a stress situation which is highly detrimental to learning and intrinsic motivation (Skowronek 1970). They will inform students about their comprehension and memory, as well as about their ability to apply methods and knowledge to problem solving. At the same time teachers will receive informational the effectiveness of their teaching. This

latter function, together with the different forms of student participation in a course, requires, in addition to the usual final examination, a continuous assessment of the progress of knowledge and qualifications during the course.

3.3.5 *Compactness of Course*

28. While the organizational advantages of compact courses do not need to be stressed, there are still uncertainties as to their effect on learning. There are very few reports on experiments with contact- courses. One of them is the experiment conducted at the "Pädagogische Hochschule Heidelberg" this year, which produced very positive results.[6] Some 500 students who had participated in one out of eight courses organized according to the "Baukasten principles" listed "compact organization" as the main advantage of this system of study.
29. It is obvious that compact courses, when organized in a liberal way, i.e. leaving enough time at the disposal of small groups and individual students, produce less interruptions, distractions and stress than today's curricula; this means that they leave more room for deeper penetration and for creativity and, because of higher cohesion, increase the chances of group learning and co-operation. As to their impact on long-term memory, it is unknown whether compact courses have a positive or negative effect. This could depend on the difficulty of the subject matter. Permanent concentration on highly abstract matters probably affects memory as well as intrinsic motivation. In this case a variety of methods should be applied or the course should be given with half-intensity, parallel to practical courses (labs: The general experience in a few compact courses indicates a considerable increase of motivation.)

3.3.6 *Small Groups and Tutoring*

30. We hold that work in small groups is a prerequisite for learning by discovery in more complex matters; it is also important for the activation and self-determination of the learning process. We expect from it a highly motivating effect and the improvement of training for communicative, social and behavioural qualifications. Yet it must be realized that these positive effects do not occur automatically; they depend on group qualities such as cohesion, affective and communication structures, norms and leadership. Those qualities are, in turn, dependent upon variables such as group size, group experience, membership structure, personality characteristics, duration and compactness of work, and nature of tasks assigned to the group.

---

[6] The results were later published in: *Hochschuldidaktischer Versuch an der PH Heidelberg*, Reihe: Hochschuldidaktisehe Versuche in der BRD, E. Meyer, ed. (Stuttgart, 1971/1972).

31. It is obvious that the membership structure will be strongly affected by the degree to which the curriculum is open, by relatively free access and permeability. Groups will be more heterogeneous with regard to age, years of study, major disciplines, and vocational experience of their members. Small group research gives almost no clues as to the effects of the various forms of heterogeneity of groups on their effectiveness. Only experience will show whether heterogeneity will represent, to a certain degree at least, a stimulating factor improving the ability to communicate or whether it would be preferable in this respect to form deliberately homogeneous groups. But whatever the results of these experiments, it seems unquestionable that small groups represent still the best way of integrating students of lower qualification or lesser experience, especially when compared to mass lectures or individual studies. Tutors or more qualified students can initiate and organize for their benefit additional exercises or provide complementary information which will enable such students to keep up with the group.

3.3.7 Learning by Discovery

32. Learning by discovery presupposes a certain interest and willingness to look behind scientifically established results, propositions and theories. Short-cycle higher education has never provided this approach, but has shown only an— instrumental interest in science and research. This attitude cannot be maintained because it destroys the spirit of enquiry in the student, which has to be developed by cognitive dissonance.[7] Only when all students have the right to raise questions will differences between scientific and other ways of study disappear.
33. The introduction of learning by discovery in short-cycle and vocationally oriented courses of studies carries with it also certain problems, because it is opposed to the tradition educational philosophy of 'Ausbildung', which implies high effectiveness of learning and a somewhat authoritarian way of teaching. These attitudes must give way to a full autonomy of students, who are themselves responsible for finding the optimal combination between activity and passivity, enquiry and pure learning.
34. Finally, students must get used to the need and importance of internal motivations which have to replace the former external motivations and help to overcome frustrations arising from problems implied in enquiry learning and in group co-operation. The growth of intrinsic motivation is, on the other hand, helped by the method of problem oriented studies and by learning by discovery.

---

[7] H. Skowronek: *op. cit.* 122–126.

3.3.8 Compulsory Knowledge

35. The concept of minimum knowledge mentioned in paragraph 13 is based on the recognition that knowledge is not a goal in itself, but that it provides means for the identification of correlations and of problems, that it facilitates co-operation with other disciplines, and the finding of solutions. Knowledge also allows transfer of concepts from one problem area (theoretical or practical) to another. This is indispensable, due to rapid change of scientific knowledge, which may be illustrated by the fact that, for example, half of the knowledge acquired in physics or medicine is obsolete after 6 or 8–10 years (Bruno Fritsch 1971). If transferability becomes a principle of higher education, knowledge taught and learned should be selected according to the theory which recognizes that transferability increases with the degree of generalization of knowledge (Klausmeier 1969). Such generalized scientific knowledge implies emphasis on structures, theories, principles, concepts and methods. Vocationally oriented courses should also, in addition to generalized knowledge, provide practical knowledge in the relevant disciplines (such as sociology and psychology for social workers). The difference between a vocational and a scientific course of studies will then consist of a somewhat smaller number of courses offering generalized knowledge, but more opportunities to apply (transfer) this knowledge to problem oriented or practical studies. The gaps of a short-cycle graduate in either theory or application can be compensated by a second cycle of studies or by permanent education.

3.3.9 New Information and Communication Techniques

36. The use of new technical media in higher education has both positive and negative implications for curricula and the learning process. One of the curricular advantages is the possibility of storing, recalling, or changing and adapting information whenever needed. A disadvantage of technical media (with the exception of learning machines and computers) is their one-way communication from the medium to the student. Consequently, there is no possibility to repeat questions or answers, no feedback for the teacher and no certainty about what the recipient actually perceived and understood. The perception is different because of interferences, varying expectations and differences in previous knowledge, as well as because of selective perception. Another problem arises with a heterogeneous group of recipients and their differences in the pace of understanding and learning. In this case, individually adaptable media should be chosen or repetitions made possible. Motivational impact of technical media is, as yet, almost unknown and seems highly dependent on the way of presentation. Various experiences with technical

media have led to a certain number of findings concerning their application in Germany:

- technical media should be part of a functionally differentiated system of learning media;
- auditive and visual aids should not be autonomous, nor should they play the leading role, but they must be closely integrated with other media (particularly with group discussions) and complement one another (Dohmen 1970);
- the functions likely to be performed by technical media are information, impulse and stimulation, control of knowledge and development of visual memory; functions not likely to be performed are development of intrinsic motivation, stimulation of creativity and change of opinions;
- the integration of technical media into the curriculum must follow different models in first-cycle study, second-cycle study and in permanent education.

37. More experience and research is necessary in order to evaluate the cognitive and affective impact of technical media on the student. Finally, more attention should be paid to the use of technical media in vocational courses.

3.3.10 Permanent Education

38. Permanent education has the advantage of bringing students with vocational experience into regular courses. Both vocationally and theoretically oriented students benefit from receiving direct information about occupational requirements, problems, and situation concerning the profession which they desire to enter (or related professions; This is an important informal complement to the formal advisory system, which facilitates decisions on courses and objectives of study. "Adult students", on the other hand, benefit from their close contact with the new student generation. It not only helps them to understand new research developments, changes in terminology, and values, of a particular science, but also to understand better their young colleagues at work (or even their children). The latter point was particularly emphasized by teachers and supervisors who participated in the already mentioned Heidelberg experiment. Nearly all of them objected to a separate further education for adults.

39. Such positive effects will, of course, diminish with an increasing social and educational distance between participants of a single course. There is no room here for discussing problems of heterogeneous learning groups, which has been taken up elsewhere.[8] It is still unknown where or whether we can find an optimum degree of homogeneity. We must leave it to the experience with

---

[8] I. Kehler, G. Kellner, E. v. Weizsäcker, *op.cit.*: 47 ff.

integrated courses and groups. Regulations are always possible by putting up course requirements.

40. A high percentage of permanent education will most probably be provided through technical media and written materials, complemented by group work. With regard to this aspect of permanent education, integration will be mainly on the institutional level.

41. Our discussion of various curricular implications of the BGHS model revealed that the majority of effects and problems cannot be predicted or solved by theoretical means, nor can they be derived from a large pool of experiments. But as long as no experiments are started, a discussion about the integration of short- and long-cycle courses of studies remains entirely academic.

42. In Germany the legal basis for the "integrierte Gesamthochschule" will soon be created, but the principles guiding its realization are rather different from the liberal, flexible and student-centred concepts presented in this paper. There is a trend bade to reglementation, to uniformity and standardization, to rigid courses of studies and frequent selective processes. This danger of regression calls for experiments that could demonstrate the superiority of an integrated post-secondary education as outlined here.

References

Bayerischer Rundfunk (Ed.), 1970: *Telekolleg, Wissenschaft-liche Begleituntersuchung*, No. 2 (Munich: Bayerischer Rundfunk): 22–23.

Bruno Fritsch, 1971: "Gesellschaftsentwicklung und Bildungssystem", in: *Neue Zürcher Zeitung*, Nr. 304, 4.7.1971: 37.

Dohmen, G., 1970: "25 Thesen zum Fernstudium im Medien-verbund", Paper Presented at a Conference on "Study without Lecture Hall" in Gummersbach, 1970.

Klausmeier, E.J.: Davies, J.K., 1969: "Transfer of Learning", in: *Encyclopedia of Educational Research*, 4th ed., R.L. Ebel, ed., 1969: 14–89.

Skowronek, H., 1970: *Psychologische Grundlagen einer Didaktik der Denkerziehung* (Hannover, 1970): 143.

Weizsäcker, Dohmen, Jüchter, et al. 1970: *Baukasten gegen Systemzwänge, Der Weizsäcker Hochschulplan* (Munich: Pieper) Revised ed., 1971.

Ernst Ulrich von Weizsäcker as Dean of the Bren School of Environmental Science and Management, University of California, Santa Barbara, with three of his students. *Source* James Badham. Copyright granted to Ernrst von Weizsäcker in 2007

Chapter 4
Contagious Knowledge: Contagion as a Quality Criterion for Disciplinary and Interdisciplinary Science

4.1 Introduction

One of the most intriguing statements about quality in the sciences is attributed to Max Delbrück: "When you are fifty and you still understand your pupils—then you don't have good pupils." Let us assume that this was not a melancholic gerontological statement—since Delbück remained highly successful in his later years, even after he had won the Nobel Prize! Then his statement either confirms the ubiquitous complaint about overspecialization in the sciences, or it does indeed convey a message about the nature of scientific quality.[1] The latter is what I believe.

The statement could then read (admittedly less humorously): If you are a good scientist, your results will generate unexpected new results and your teaching and supervision will elicit creative thinking in your students. And this will go on with such rapidity that after a decade or two when many of your pupils have settled down in their own laboratories around the country, most of their work will be beyond your comprehension—provided they are good scientists.

Scientific quality, then, would include the propagation of ideas into the unknown by which process the high-quality ancestors of the ideas can even become obsolete. Let me call this the *evolutionary aspect of scientific quality*, owing to its resemblance with the succession of forms in a short first part of this chapter I am offering some observations concerning scientific quality. They will amount to a substantiation of the evolutionary aspect mentioned. This will lead me in the second part into a more philosophical quest for the underlying structure of 'contagious' or 'infectious' information. Having done that, I shall proceed to a classification of six different types of contagious scientific knowledge, which will

---

[1] I wish to thank my wife Christine von Weizsäcker for many talks and for very helpful comments on the draft version of this chapter. Also I owe much to the comments by my colleagues M. Anandakrishnan and Lilly Landerer at UNCSTD, and later by the participants of the Nobel Symposium, especially Professors Menon and Simon. Earlier talks with my father Carl Friedrich von Weizsäcker and Georg Picht (1913–1982) founded the basis of the thoughts forwarded in this chapter. This text was presented to the 58th Nobel Symposium 15–19 August 1983 in Oslo.

be done in the third part. Interdisciplinary research will receive special attention in this part. And in a final fourth part, I shall venture a few words about quality and responsibility in both disciplinary and inter-disciplinary research.

4.2 Scientific Success

Scientists, it is commonly said, know everything about nothing, while journalists and politicians know nothing about everything. The saying puts into the extreme what many people feel and experience. There seems to be an optimization problem, if not a dilemma. We may hope to understand a tiny part of reality, seen from a very narrow, i.e. well-defined methodological angle. But by restricting ourselves that much, we may be sacrificing all the links to the complex reality surrounding us. Good scientists usually defend the modesty in scope and the well-defined disciplinary methods by hinting at the success stories of the various sciences: The free fall in the exotic, *unreal* vacuum gave Galileo more insights into the *real* laws of motion than millions of real falling leaves could have done, and the grandiose simplicity of phage genetics, developed by Delbrück and Luria[2] helped the advancement of biology more than thousands of practice-oriented animal-breeders did over the centuries. And if the tiny piece of reality happens to be a onc-gene[3], then no knowledgeable person would deride that research as meaningless.

On the other hand, Galileo, Delbrück and the discoverers of the onc-gene are the exceptions. Much of the everyday work in scientific laboratories and in social science programmes seems to follow the philosophy of that poor man who lost his key at night and who was looking for it under the only lantern on the street; and when somebody asked him if he was sure that he lost it *there*, he said: "No, but here under the light I have at least a chance to find it."

Some exceptionally good papers are devoted to "installing more lanterns" or to "producing torches"—if you permit me to stay in the simile. To widen the range of the light, that is where the success lies. The virtue of successful science is rather not simplicity per se but the opening of new frontiers, the "endless frontiers" in Vannevar Bush's famous words (Bush et al. 1945).

There are serious efforts to quantify the stimulative success of scientific papers. The simplest method is the counting of the citations a paper receives,[4] which can

---

[2] For a historical account of phage genetics and its broad impact, see "The Festschrift for Max Delbrück", in: Cairns, J.; Stent, G.; Watson, J.D. (Eds.), 1992: *Phage and the Origins of Molecular Biology* (Cold Spring Harbor).

[3] Onc-genes are genes which can be "switched on" to transform the cell into a cancer cell. For a recent review, see Duesberg, P.H. 1983: "Retroviral Transforming Genes in Normal Cells?", in: *Nature*, 304(21 July): 219–226.

[4] The Science Citation Index lists all new citations earlier papers have received in major journals during the time period covered. The main value lies, of course, in quickly guiding specialists to all new papers which build on known existing papers. The SCI is published quarterly and in a cumulative version annually by the Institute for Scientific Information, Philadelphia. The use of

be easily done using the Science Citation Index. Not surprisingly, trail-blazing papers for which their authors were awarded the Nobel Prize, ranked highest in citations even before the award was given. The explosion of papers following certain key papers was even described successfully by epidemiological models[5]. The Nobel Prize papers represent prime examples of what we may call 'contagious' scientific knowledge. Conversely, boring, repetitive, incomprehensible or faulty papers have little effect, with the exception perhaps of the temporary limelight some faulty papers receive until they are refuted.

It has to be admitted that some systematic errors are contained in any such short-term citation or impact quantification. There are fashions, there are sterile "mutual admiration societies" and there are the vested interests in the continuation and enhancement of disciplines and sub-disciplines (Cole et al. 1981). The peer review system for both grants and publications ensures a certain inertia or self-reproduction of sub-disciplines—(remember that the peers often are over 50!). Most of these systematic errors, however, persist over a relatively short period of time only, leaving the long-term infectiousness a good criterion for good science and short-term infectiousness at least an important indicator. In any case it seems worth the effort to take a more systematic look into the nature of contagious information.

4.3 Contagious Information

Contagious or infectious processes you can find all over the place. There are contagious diseases and there are bacteria (the latter being Pasteur's scientific answer to the riddle of the former). There are certain modes of contagious behaviour, like giggling, applauding and, sometimes, hysteria. But there are also more primitive phenomena like wave propagation and, most important, the transmission to the next generation of genetic information. It was this multitude of evidence which led C. F. von Weizsäcker[6] to his formulation: "Information is only

---

(Footnote 4 continued)
the SCI for additional purposes including the evaluation of impact is discussed in: Garfield, Eugene, 1979: *Citation Indexing* (New York: Wiley). See also Porter, A.L. 1977: "Citation Analysis: Queries and Caveats", in: *Social Studies of Science*, 7: 257–267.

[5] The use of mathematical epidemiology for the spreading of scientific knowledge was introduced by Goffaan, William; Warren, Kenneth S. 1980: *Scientific Information Systems and the Principle of Selecticity* (New York: Praeger). The purpose of this study is to develop a criterion for 'quality' papers and 'quality' journals, so as to help libraries in distinguishing them from the junk.

[6] Carl Friedrich von Weizsäcker developed this notion in the context of physics but it is meant to hold also in other disciplines. See von Weizsäcker, C.F. 1971: "Materie, Energie, Information" in: *Die Einheit der Natur* (München, Hanser): 342–366, especially, 351–352. The ideas are developed further in von Weizsäcker, C.F. 1977: *Der Garten des Menschlichen*, paperback (Frankfurt: Fischer, 980): especially 139–152.

what creates information." This statement can be seen as the defining condition[7] for contagious information. It clearly departs from the famous definitions by Shannon and other authors who attributed information a mathematical value determined by the negative logarithm of the frequency-derived probabilities.[8] "Contagious information" tries to revive, on the other hand, the spirit of the pioneer days of cybernetics when it was discovered that human knowledge and technical signal transmission had something essential in common, namely information. The early enthusiasm vanished when it became apparent that Shannon's quantification would hardly help to explain knowledge or real life. Real life consists of unique events with no established frequencies from which to derive probabilities. In particular, scientific discoveries own no frequency beforehand, and human knowledge does not operate on anything like stable probabilities. It is only on the semantically low levels of signal transmission such as the level of letters that you can count on stable frequencies and, thereby, probabilities.

The difficulty, on the other hand, with the definition of contagious information is its seemingly tautological form although it would find objections only in its stronger ontological understanding[9], not in the more cautious conditional form. However, even the stronger form can be defended by hinting at 'tautological' statements in biology such as "a duck is what creates ducks", or "fit are those which survive", or "successful in evolution are those which have a successful

---

[7] The statement expresses a *condition* for contagious information. It is tempting to go a small step further and to omit the word 'only' so that the statement would read "Information is what creates information". This would be a quasi ontological *definition* of contagious information. The later parts of my paper are consistent with this ontological definition. Thus I call the statement a *defining condition* indicating thereby that I would accept saying that whatever creates information may be called information. It does not harm but it does not help much either if one distinguishes by index numbers 'parent' information from 'children', "grand children", etc. information. The notion of contagious knowledge is an old theme in philosophy. it is the central notion in the chapter "Die Bildung" in Hegel's "Phänomenologie des Geistes".

[8] Shannon, Claude E.; Weaver, Warren, 1949: *The Mathematical Theory of Communication* (Urbana: University of Illinois Press). There have been many modern reformulations of Shannon's work, especially by Brillouin, Leon, 1962: *Science and Information Theory* (New York: Braziller) who, in my view, rather confused the matter by calling information 'negentropy' (for a discussion of this point see Zucker, Francis, 1970: "Information, Entropie, Komplementarität und Zeit", in: von Weizsäcker, Ernst (Ed.): *Offene Systeme I* (Stuttgart: Klett): 35–81. However, all reformulations adhere to the idea that information should be quantified using the logarithm of the expectation probabilities.

[9] There are various approaches to the quantification of semantically meaningful levels of information, notably by Bar Hillel, Jehoshua; Carnap, Rudolf, 1953: "Semantic Information", in: *British Journal for the Philosophy of Science*, 4: 144–157. For an overview heeding especially the important work of Donald M. MacKay, see Nauta, Doede, 1972: *The Meaning of Information* (Den Haag: Mouton). Not surprisingly, none of the approaches overcomes the problem of quantifying probabilities of unique events. Unfortunately, the inapplicability of Shannon's qualification to higher levels of meaning has led to a language use restricting 'information' to meaningless levels. It is my expectation that a deeper understanding of the higher levels, spurred by brain research and the development of artificial intelligence, will lead us back to a meaningfull meaning of information.

progeny" and non-tautological definitions of ducks, fitness and success tend to miss some essentials of life: time, change and fragility.

My wife Christine and I[10] have made attempts to explore the conditions of success for contagious information. To summarize our ideas in a few words: We consider two complementary components as necessary for each information act: *novelty* and *confirmation*. If novelty is missing, there is nothing new, thus there is no event at all. Where there is no confirmation, we have chaos ('Confirmation' is a rather tricky notion which may have to be further developed; it contains such a variety of things as redundancy, physical reliability, e.g. of the channel, physical intensity of the message, existence of a code of some kind, and a few more things; 'novelty', on the other hand, is a more straightforward notion, which under the "high confirmation" conditions of no-noise communications engineering is identical with Shannon's information, but under noisy conditions also contains that noise). The highest information production is attained where there is a good balance between novelty and confirmation. The situation is qualitatively indicated in Fig. 4.1.

Any increase, jointly or independently, of the two components enlarges the potential for pragmatic information or contagion; think of a theatre play, of a learned discourse or of a holiday trip full of pleasant novelty but also confirming your sense of language, your knowledge, or your health: the impressions stimulate you to further creative activity including the creative use of the information acquired. If the process concerned was basically boring you will enjoy any change even at the expense of some confirmation; if it was basically confusing you will welcome repetitions, redundancy or other confirmation.

The figure is a representation in principle of the optimization problem referred to in the beginning: knowing "everything about nothing" means too much confirmation, knowing nothing about everything means too much novelty. Good, contagious science skilfully keeps the balance. If someone suspects having discovered something highly surprising (high up on the novelty axis in Fig. 4.1) he will have to assemble much confirmation to attain a balance, i.e. to reach the maximum of pragmatic or contagious information.

The picture is over-simplified so far. There are aspects of reality, where you wouldn't want disturbance or novelty at all. Introducing into the alphabet new letters would be improper even for the most avant-garde poet: and our body strictly fights the intrusion of alien molecules. Less strictness is applied on higher organizational levels such as words or combinations of nucleotides in the DNA, but

---

[10] von Weizsäcker, Ernst; von Weizsäcker, Christine, 1972: "Wiederaufnahme der begrifflichen Frage: Was ist Information?", in: *Nova Acta Leopoldina*, 206: 535–555. Reprinted with small modifications in von Weizsäcker, Ernst (Ed.): *Offene Systeme I*, loc. cit. In both papers novelty (Erstmaligkeit) and confirmation (Bestätigung) were conceived as being representable on the same axis, if in opposite directions. This has meanwhile been corrected in favour of a two-dimensional plane with novelty and confirmation increasing in perpendicular directions and pragmatic information represented on a third axis, thus demanding a three-dimensional representation, as in Fig. 4.1. See von Weizsäcker (1984).

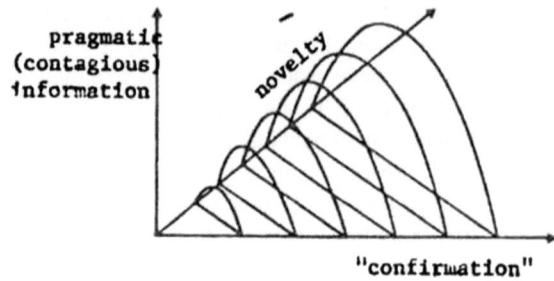

Fig. 4.1 Qualitative three-dimensional graph for the growth of contagious information in dependence of novelty and confirmation

still, novelty is kept at very low levels. It is only on the highest hierarchical levels available, such as meaningful speech, animal behaviour or population genetics that novelty and confirmation ought to be equilibrated (as in Fig. 4.1) to ensure maximum information production.

Yet, we have to be very careful using the term "hierarchical levels". In certain processes, notably in science, the *hierarchy seems inverted* in that 'lower' levels may as yet be unexplored so that all the scientific curiosity focusses on them and all the important subjective novelty concerns them.[11] Think of the deciphering of the genetic code or of an antique language. In those processes the more complex edifices of scientific knowledge serve as the confirmation needed for taming and embedding the novelty-rich empirical findings. Similarly, to focus attention on an individual in a group or on a particular event among many others amounts to making this individual or that event the novelty-bearing 'highest' level.

Successful information transmission cannot occur where there is no suitable *code*, no *channel* or no *vector*. While the code and the channel belong mainly to the domain of 'confirmation', the vectors normally are exposed to novelty and they produce novelty. The vectors cannot be meaningfully distinguished from the *receivers*. In fact, if there is a receiver who is not a vector in that he only receives and absorbs the information without forwarding anything and without even storing it, he 'kills' it (although it may be alive somewhere else). The vectors, or receivers, will not normally understand the full message and their transmitting capacity will be limited. In addition they may have reason to highlight a special part of it and to actively disregard another. This is the novelty they produce.

Thus, the information is likely to change its meaning as it spreads, and grows, just as Max Delbrück would have expected. The proper key and lock matching between a message and its receiver is crucial for contagion, for the success of the message. The locked door, we may also say, 'understands' the fitting key. A receiver in whom the message matches with his previous knowledge and in

---

[11] The term "subjective novelty" indicates that the creation of pragmatic information out of confirmed novelty is not restricted to things or objective phenomena. To gain knowledge about things or objective phenomena also requires a balance of novelty and confirmation.

whom it causes the satisfaction of understanding is likely to carry the message along—sooner or later, and perhaps in unexpected contexts, thereby constituting the success of the message. In science the lock could also be a problem and the key an answer solving the problem.

A scientific theory, for instance, which serves beautifully to solve a given problem—that is considered a success. We see this as analogous to the hummingbird's answer to the deep calyx of flowers or any other successful adaptation to an ecological challenge.

Hummingbirds cannot thrive in an environment without flowers. Scientific knowledge which might be potentially contagious, but is deprived of an audience cannot thrive either. In an illiterate a completely untechnical environment even the genius of Thomas Alva Edison could not have matured. Thus, to help create *an environment* rich in codes, channels and vectors may be equally important for the advancement of sophisticated knowledge as to create such sophisticated knowledge itself; just as some species, notably bees, beavers and humans, celebrated the greatest adaptive successes by purposefully engineering the environment to their suiting.

Let us briefly consider at this point the philosophical dimension of the key-and lock concept[12]. According to the concept, *scientific answers are not dead rocks in a landscape of truth*; rather they would be fitting keys in a landscape of locked doors. Yet, rocks of truth and fitting keys have something important in common: Both refer only to the *confirmation* part of contagious information, or the *correctness* aspect of scientific success. In biology the fitting key refers to the short-term survival or fitness aspect only. But yester day's correctness or survival may not suffice, if the problem or the biological challenge has undergone any change since. In consequence we seem to be forced to add to the fitting keys notion a complementary component meeting the challenges of *novelty*. This component will have to account also for the innovative aspect of both scientific and biological success; and that component will specifically have to deal with *errors*, *misfortune* and *destruction*.

The study of biological and technical systems has led Christine von Weizsäcker to introduce the term "error friendliness" for this component[13]. To limit the effects of errors and of external attacks, to cultivate a learning-from-errors capacity, and even to harness occurring irregularities for innovation (as you do in "productive misunderstanding" and as species do with mutations), those are the main features of this capacity. Error-friendliness stands complementary to short-term fitness. Only together the two components warrant long-term success.

---

[12] Great philosophers, notably Hegel, Husserl and Whitehead have forwarded similar thoughts. Most specifically, the notion of truth as lock-opening keys was proposed by von Weizsäcker, Carl Friedrich, 1971: "Modelle des Gesunden und Kranken, Guten und Bösen, Wahren und Falschen", in: *Die Einheit der Natur* (München: Hanser): 320–341.

[13] von Weizsäcker, Christine, von Weizsäcker, Ernst, 1984. 'Fehlerfreundlichkeit' in: Kornwachs, Klaus (Ed.): *Offenheit, Zeitlichkeit, Komplexität* (Frankfurt: Campus). This chapter explains why such disparate notions as physiological resilience, locomotion, mortality, mutations and the gene pool, ecological resilience and geographical isolation can all be seen as factors providing "error-friendliness".

The dinosaurs overdid the short-term fitness component and lacked the error-friendliness to survive the geophysical turmoil of the late Cretacean time. Conversely, those inconspicuous little warm-blooded mammals of the time quickly adapted and survived; and they were still versatile enough to give rise to a whole phylum of different species. All mammals of today, rats and bats, and cows and whales, not only their closest surviving relatives, the insectivores (e.g. the shrews) bear witness of the early mammals' success; (although from the point of view of a shrew, whales or cows are certainly appalling errors!).

So far we have treated the contagion as if it was undoubtedly welcome. This is not so. Successful contagion is normally *ambivalent* and sometimes even *destructive*. It may be good for one and bad for another one. Cancerous cells are highly successful on their own account but they may destroy the body to which they belong (which amounts to self-destruction of the cancer cells, proving thereby the ambivalence for them of their unchallenged success). Think further of mass hysteria or of the criminal abuse of knowledge. We get only little comfort from the fact that such destructive processes usually lead to the natural end of their self-destruction, since before that end they can inflict inexpressible suffering on innocents. Yet, from the theoretical point of view, it has to be noted that destructive parasitic or 'selfish' information is not sustainable.[14] The most successful type of information is 'symbiotic': It benefits and enriches its receivers or its hosts, who in turn will try to get more of this treat; they will save it and enhance it and they will flourish in doing so.

4.4 Contagious Science

After this lengthy detour into the philosophy of biology, contagion and information let me now return to the discussion of scientific success. Let us assume that success and quality in science should indeed be measured by its infectiousness. According to the philosophy of contagious information such measurement would have to consider a variety of receivers, just as the Cretacean shrew's information has been received (and adapted and carried along) by many major groups of mammals. By distinguishing major groups of receivers we shall arrive at a *classification* of the scientific endeavour into six groups:

---

[14] In this context it should be noted that the fashionable concept of DNA or of the genes being 'parasitic' to the 'hosts', i.e. the organisms to which they belong (Dawkins, Richard, 1976: *The Selfish Gene* (Oxford: Oxford University Press). Dawkins, Richard, 1982: *The Gene as the Unit of Selection* (Oxford: Oxford University Press). Crick, Francis H.C.; Orgel, Leslie, 1980: "Selfish DNA: The Ultimate Parasite", in: *Nature*, 284: 604–607. Doolittle, W. Ford; Sapienza, Carmen, 1980: "Selfish Genes, the Phenotype Paradigm and Genome Evolution", in: *Nature*, 284: 601–603), is, of course, limited by the tolerance of the 'hosts'. Moreover, (as argued in the text), symbionts which effectively help the survival and evolution of their hosts, strive better than neutral or destructive parasites; and this naturally holds also on the level of DNA and the genes.

1. *First*, and most prominent for scientists, there is the *self-infection* in the sciences. One good paper gives rise to the next, and sometimes to a whole epidemic of papers.

   Let us furthermore remember the need for an environment rich in codes, channels and vectors. This has to be created and sustained, by *science education* and *science writing* and otherwise reaching people outside the existing scientific community; in the concept of contagious information this would be no less important than the primary production. Thus the present work of Abdus Salam[15] and his colleagues at Triest, opening first class physics training to students from the third world, may be viewed, a century later, as a greater contribution to self-infection of science than the scientific work which earned him the Nobel price: think of one or two Edisons or Einsteins who otherwise would not have found a conducive environment.

2. *Second*, there is the technological infection. The scientific information has to make a big leap like that from insectivores to bats, to prove its usefulness in a totally different medium, namely under competitive commercial conditions. Success in the new medium is in no sense inferior to success in the old one, provided it is not a dead end road. If the technology helps to feed people or to enable them to help themselves, it is not a dead end road and it is symbiotic information in the best sense And, if the applied scientist loses track of the branching success of his work, he may remember Max Delbrück who did not complain about the analogous phenomenon in the pure sciences.

3. Let me now come to the *third* group of contagious sciences. It is the cross-fertilization in what good scientist refer to as inter-disciplinarity, e.g. in bio-physics. To distinguish it from other forms of interdisciplinarity which I shall mention later, let me call it *hybrid or methodological interdisciplinarity*. Best known are perhaps the myriads of physicists who have swarmed out into biology to crack the nuts for which the biologists somehow had not the right jaws. The themes ranged from the aerodynamics of albatrosses to the structure of DNA. In a similar way physics has thoroughly transformed chemistry, astronomy, and the earth sciences; and mathematical methods, increasingly strengthened by computer capacity, have been applied to economics, law, linguistics, psychology and sociology. The Annals of the Nobel prizes for chemistry, medicine and physiology and, more recently, economics, record many success stories of that kind. It is worth mentioning also that technological progress continuously inspired or infected scientific methodology, thereby confirming the 'symbiotic' relations between science and technology.[16] In a particularly striking way modern communication technology including

---

[15] Mentioning the merits of individuals may be found improper in the context of this chapter. On the other hand, the otherwise fairly abstract reasoning seems to require some illustration.

[16] The 'symbiotio' relations between science and technology were highlighted at the Nobel Symposium by M. G. K. Menon. See his contribution in the same volume, where this chapter was first published.

computer conferencing has become an important tool for both research and education, in both cases enhancing the contagion of infectious ideas.[17]
Hybrid or methodological interdisciplinarity has gained full recognition among scientists and science policy makers. Only the traditional intuitive masters of the receiving sciences sometimes became nostalgic and did not find the invading information very symbiotic. They saw that something was lost, and they could be right.

4. In trying to understand what it is that may have been lost, I am guided to the *fourth* group of contagion in the sciences. The phenomenon I am dealing with may be called *complex contagion or complex interdisciplinarity*. The study of an ecosystem including human settlements cannot be reduced to biological, physical and chemical problems. And European Studies means more than just languages, trade and institutions.

Many modern 'disciplines' belong almost entirely to this kind of complex interdisciplinarity, like area studies, agriculture, energy systems research[18] or psychosomatic medicine. But also the age-old fundamental questions of the nature of time[19] or the nature of the self cannot be satisfactorily tackled by any one of the three first types of scientific endeavours mentioned so far. Complex interdisciplinarity always seems to cross courageously the barrier between the Two Cultures identified by C. P. Snow.

No doubt, the ambition of complex interdisciplinarity is breath-taking. Universities don't like this. They avoid such complexity which is difficult to transcribe into multiple-choice or other objective tests. They manage to present

---

[17] See, e.g. Hiltz, R. 1982: "Experiments and Experiences with Computerized Conferencing", in: Landed, et al. (Eds.): *Emerging Office Systems* (Norwood, N.J.: Ablex), cited in *WAAS-MIRCEN-IFIAS* (Stockholm: Aug 1983); A Dispersed Conference on Biconversion of Ligfto-cellulose for Fuel, Fodder and Food, Needed for Rural Development in Poor Countries, 12–17 December 1983. More information can be obtained from Prof. Carl-Göran Heden, Bacteriological Institute, Karolinska Institute, S-l0401, Sweden.

[18] Complex energy research may perhaps deserve the palm for breaking the ground in complex interdisciplinarity. It began. in the 1950s with an interdisciplinary development and assessment of nuclear energy. Energy demand and technology assessment led Wolf Häfele to his concept of 'hypotheticality' (the insight that planning cannot wait until all contingencies are removed and therefore has to rely in part on unproven hypotheses; Häfele, W. 1972: *Hypotheticality and the New Challenges—The Pathfinder Role of Nuclear Energy* (Laxenburg, Austria: IIASA). Later it became evident that the non-nuclear path involved even more interdisciplinary reasoning (Lovins, Amory, 1977: *Soft Energy Paths* (Pelican Books). And today the food-energy-nexus appears to be one of the most promising research themes with practical relevance (see note 22).

[19] An impressive series of multidisciplinary, sometimes interdisciplinary volumes has been published by the International Society for the Study of Time. Easiest to obtain and to read may be: Fraser, Julius T. (Ed.), 1981: *The Voices of Time* (Amherst: University of Massachusetts Press), 2nd edn. The philosophical dimensions of time and its connections with responsibility, reason, peace and human ecology are discussed in a very challenging way by Georg Picht in his two volumes: *Hier und Jetzt—Philosophieren nach Auschwitz und Hiroshima* (Stuttgart: Klett-Cotta, 1980), and earlier by Klaus Müller, A.M. 1972: *Die präparierte Zeit* (Stuttgart).

even agriculture, area studies, etc. the dissected way. Regarding research only a few universities, most prominently the University of Sussex, England, are systematically exploiting the exuberance of fascinating scientific problems requiring complex interdisciplinarity.[20] Outside the universities there is more abundance of complex interdisciplinarity. The *International Federation of Institutes for Advanced Studies* (IFIAS)[21] connects internationally institutions devoted to systematic complex inter-disciplinary research at high levels of excellence. And the *United Nations University* (UNU), which is not a teaching university in the ordinary sense, is involved with a great variety of such projects, not the least the newly launched study of the "food-energy-nexus"[22].

It should be emphasized that complex interdisciplinarity cannot live without *team work* (this is also true, of course, for most other forms of research). Inside the teams a constant effort of translation from one jargon into another has to take place. Every individual has to find a proper balance between cultivating his own discipline (to remain competent there) and plunging into the common problem. Highly perceptive receivers are needed to keep the criss-crossing information alive and meaningful. A quasi artistic ability to perceive or to smell connections is of use. Just as in the old soft disciplines with their intuitive masters. Complex interdisciplinarity may offer a revival for such capacities, but certainly not at the expense of the methodological quality introduced by the hard sciences.

5. After we have reached the level of complex interdisciplinarity, you may wonder what would be the fifth and sixth categories of the scientific endeavour. Well, I am now turning to realities outside the ivory tower and also beyond the sheer one-way application with which we have dealt under the second category.

The *fifth* category which I call *practice-oriented interdisciplinarity*, comprises complex scientific and technological knowledge and action in the real world: The WHO Programme to eradicate the small-pox in the world; research-based agricultural innovation in a tropical African region under the given economic and political conditions; or the Kerala people's science movement, which does not wait until the key actors have college degrees.

---

[20] In a much more modest manner the young West German University of Kassel (Gesamthochschule) has started to establish interdisciplinary research centres and working groups. See, e.g. *Bericht des Präsidenten*, Gesamthochschule Kassel, 1980.

[21] IFIAS has so far 28 members Institutes with specialities ranging from international relations to livestock management. It conducts eight interdisciplinary programmes and projects such as "Land and Water Resources for a Sustainable Biomass Production". The secretariat's address is: Ulriksdal Slott, S-17171 Solna, Sweden (Dr. Sam Nilsson).

[22] Information on the programmes of the UNU can be obtained from: UNU, Toho Seimei Building, 1–15, Shibuya 2-Chome, Shibuya-ku Tokyo 150, Japan. A condensed account of the interdisciplinary significance of the UNU's programmes is given by Parpia, H.A.B. 1983: "An Approach to Training and Institutional Development Needs for Utilization of Emerging and Traditional Technologies", in: von Weizsäcker, Ernst; Swaminathan, M.S.; Lemma, Aklilu (Eds.): *New Frontiers in Technology Application* (Dun Laoghaire, Dublin: Tycooly): 216–220. Correspondence concerning the Food-Energy-Nexus programme should be directed to Prof. Ignacy Sachs, CIRED, 54 Boulevard Raspail, 75270 Paris, CEDEX 06.

One of the finest examples of successful, practice-oriented inter-disciplinarity is the Green Revolution in India. After the breeding of high- yield crops by Norman Borlaug, B.P. Pal and others, one of the main problems was their broad scale application. Campaigns to introduce the new crops had to bring to a match the technological and procedural changes require and the readiness for change in the rural population. One of the great managers of this important step was M.S. Swaminathan, then Director of the Indian Agricultural Research Institute and an eminent scientist himself, who strongly stressed the active involvement even of the illiterate farmers[23]. At the end, India, which in 1967 had been labelled as "can't be saved" by the authors of "Famine 1975" (Paddock/Paddock 1975) became more or less self-sufficient regarding food, although many questions of distribution remain to be solved.

The examples given are taken from developing countries, where there is at least a broad basic agreement on the goals, namely relief from misery and the improvement of livelihood. In industrialized countries the tasks are somewhat more complicated and they would involve more loaded controversies.

In particular, there can be strong and well-founded opinions against the introduction of certain new technologies, as can be seen in the nuclear energy debate. On the other hand, to take part in such a debate (not necessarily to take sides immediately) may be a very appropriate way for a scientist or engineer to engage in practice-oriented interdisciplinarity.

Turning now to the *flow of contagious information* we observe that in the case of practice-oriented interdisciplinarity the flow goes in both directions, especially 'upstream' from the people concerned to the specialist, forcing him to adapt his perception of the problem. (Scientists need not be disturbed about this, since molecules and frogs also force us to change our perception from time to time.) The success of such collective work with contagious information naturally would lie in the success of the programme, like the actual eradication of the small-pox. Moreover, any success story, any partial, success and any failure will be a fascinating subject for complex interdisciplinary research, which may prove its contagion by leading to practical corrections.

Practice-oriented interdisciplinarity is by no means restricted to megaprojects. The curing of an individual patient with multiple ailments, the continuously changing design of a backyard vegetable garden, and the scientific advice for the building-up of a high tech portfolio can Just as well be considered activities of practice-oriented interdisciplinarity. And again the quality will be rated by measurable success.

6. Now that I have mentioned five classes of contagious scientific information I sincerely hope that you still feel uncomfortable. Something important is still

---

[23] For an account of M.S. Swaminathan's impressive life work, see Ramanujam et al. (1980). The Green Revolution in India has been widely described. A well documented assessment also addressing the shortcomings and the economic considerations on the farm level is given by Dasgupta, Biplab, 1977: *Agrarian Change and the New Technology in India* (Geneva: UN Research Institute for Social Development).

missing. The whole picture looks so suspiciously harmonious, doesn't it? Scientific information was seen as contagious either within the discipline or beyond, and the worst that seemed possible was non-contagious information or, what would amount to the same, deaf and blind receivers. Unfortunately, this is not so. The worst case is *destructive information*. This is the *sixth* category of contagious information. It is real, just as the criminal and military misuses of scientific knowledge are real. This cruel reality, on the other hand, corroborates the validity of the contagious information concept: A dead rock in a landscape of truth would not do any harm. But it would not be alive. The price of living truth is danger. Let us face the danger.

There are scientists employed by the defence ministries in East and West (alphabetical order) who are studying purely scientific papers for their potential usefulness for misuse. They do it, so they would say, only to explore the hypothetical misuse by the hypothetical enemy. But to predict his sinister successes a bit more precisely they regret having to do some experimentation into the sinister direction. And to prevent the enemy from misusing the results, the experiments have to be kept secret, of course. And the more destructive the detected potential, the more excited the superiors become (solely because of the potential destructive potential of the enemy, or course). My stomach revolts I find it hard to face the danger. Let me at this point only observe that mistrust, which is the basis of that pathological research, is also contagious.[24]

Destructive scientific information is not the privilege of the military and criminal sectors. By multiplying the dimension of impacts a new order of magnitude for man-made civilian disasters was attained. By intensifying communication (precisely using the contagious property of information) the spreading of destructive information is also facilitated. And by crashing all geographical, cultural and taboo barriers our scientific-technological- economic quasi-culture is more and more sacrificing the inherited 'error-friendliness'.

Here we are at the heart of the full ambivalence of contagious information. By their very success and by their reaching out into practical applications, the contagious sciences render our world more vulnerable every year.

4.5 Quality and Responsibility

The notion of contagiousness as an important criterion for the sciences has led us to reconsider the 'ecology' of the landscape of truth. From a better understanding of this landscape I would not only expect clues for scientific quality but also for the

---

[24] This is not to say that mistrust cannot be justified. But in assessing the costs and benefits the contagiousness has to be taken into account.

responsibility questions associated with the destructive potential of science and technology.²⁵

Sir Alexander Fleming discovered penicillin which healed millions of people. This made his discovery immensely more 'real' than thousands of other discoveries which went into the archives only. The reality of Fleming's discovery lies in the case histories of millions of patients treated with penicillin, and even of comparable patients who did not receive it. This rich and complex reality forms the better part of the scientific truth around that drug. And should one day all germs be resistant to penicillin, the truth would be that the drug would soon disappear from the shelves of doctors and pharmacies and a little later from the practitioner's textbooks; and it would end up as a chapter of medical history.

Once again, if we accept contagion as the prime criterion for scientific success, we are bound to accept also the notion that the truth of a discovery includes as an essential part its effects. Let us now assume that scientific quality is valued by the success in searching for truth; then we would arrive at a fairly surprising conclusion: As long as the effects of a discovery are completely unknown, the truth-finding quality of the respective research would still be low. Or, to put it positively: to study unknown effects of science and technology²⁶, and to study methods of handling them should deserve the same type of glory of truth-finding as basic science itself. Even without public glory there are quite a few outstanding scientists, who have devoted precious years of their lives to the study of ways and means to control the military misuse of scientific knowledge; among them are the Symposium participants Carl Göran Heden and Ivan Malek.

As already indicated in the preceding chapter, the quality line in our new landscape of *truth would not be drawn between basic science* and practical application but between *contagious work and sterile work*. The discovery of a surprising technological option or of a surprising risk-reducing mechanism (including even financial and administrative components) should under normal circumstances be just as contagious as a striking scientific discovery. And even the skilful exploration of applications under a variety of conditions and the arousing of potential users can be highly contagious in the best sense; and it can be of the

---

[25] Reference is made again to the philosophy of Georg Picht (see note 21), including his earlier work, collected in the volume *Wahrheit, Vernunft, Verantwortung* (Stuttgart: Klett, 1969). Picht summarises his time-dependent notion of truth and its connections with responsibility in one sentence: "in the domain of a truth which is no longer conceived metaphysically but from the nature of time, the inherent possibility for reason can only be founded in the responsibility of man for his future history". He specifically rejects the metaphysical claim of recognizing timeless truths.

[26] This is normally called "Technology Assessment". It has been institutionalized in many countries, most prominently perhaps in the U.S.A. with the Congressional *Office for Technology Assessment*, Washington, D.C. The dimensions of technology assessment were laid out in the OECD publication by Hetman, François, 1973: *Society and the Assessment of Technology* (Paris: OECD). A recent survey of and some teaching approaches to technology assessment are given by Liao, Thomas T.; Darby, William P. 1982: "Technology Assessment", in: *Bulletin of Science, Technology and Society*, 2: 583–624.

4.5 Quality and Responsibility

highest quality. (For scientists it may be comforting to remember that striking discoveries are rare exception any way and to learn on the other hand, that working in interdisciplinary teams and working in practice with ordinary people and experiencing their gratitude, is normally considered, by those who do it, as highly rewarding.)

Nevertheless, many good scientists will still feel rather irritated by the prospect of that new time-dependent landscape. They will perhaps not plainly contradict; but they will remind us of the bad experiences almost each generation in modern history has made with movements which started with some comprehensive interpretation of reality, (mostly political, sometimes religious), and which ended up in suppressing people, especially free thinkers. Ideology triumphed over objectivity in the Nazi period and under Stalin, in China's Cultural Revolution and in some Christian and Islamic societies. Also the students' movement in the late 1960s, and the current 'green' movement are reproached for tampering with objectivity and truth. Conversely, they reproach the scientific establishment for prostituting itself to money and the state.

Again and again have scientists given into temptations (and be it only to win the favour of their students). Scientists, after all, are human beings. On the other hand, they became heroes of science when they did not give in even to the strongest pressure, like Giordano Bruno, Robert Oppenheimer in the McCarthy time, or .the Russian geneticist Vavilov who lost his life contradicting Lyssenko.[27]

So much for the historical reasons to be very cautious with the interdisciplinary, comprehensive views of reality. But there are also important *philosophical reasons* to be cautious, especially against elevating social realities to the level of 'truth'. (Actually, I have, not done that; I only said that truth-*finding* should include the study of unknown effects.) Truth itself and all its connotations refer to something stable and reliable. Scientific truth refers to something testable, something independent from the value of system of the observer, something with so to speak 100$ confirmation. We will have to make sure indeed that in building the new landscape the undisputed merits of objectivity will not be sacrificed.

I have tried not to belittle the magnitude of the problem: Using the universal reality of contagious information we have arrived at a very comprehensive view of the sciences. This view was fostered by the hope to get a grasp not only of the 'anatomy' of the sciences, but also of its 'physiology' and, by leaving the ivory tower, also of its 'ecology' including its perilous potential[28]. On the other hand, we found strong historical and philosophical reasons against shaking the bedrock of truth. In the concluding part of my chapter I am now trying to offer a few humble

---

[27] For a more thorough account of anti-scientific, anti-intellectual movements and events see J. Ben David, in the volume, where this chapter was first published.

[28] The problematique is well described by Meyer-Abich (1979). See also von Weizsäcker, C.F. 1977: *Der Garten des Menschlichen*, loc. cit.: 45–65 and 66–77.

thoughts on how to reconcile the comprehensive view with the traditional cautious view of scientists who warn against intermingling with politics.

Let us remember that complex systems invariably consist of many hierarchical levels (neither rigid nor unequivocal, as said before). Our physical and social realities give ample evidence of this. Practice-oriented interdisciplinarity (which has posed us the pressing problems of quality and responsibility) mirrors this multi-level situation. Agricultural reform involves molecular facts, plant breeding results, soil properties, machinery, the farmers and their families, markets, extension, administration, to name only the major levels. All levels have to be reflected by the success-oriented actors (farmers, scientists, administrators) even if their actual deeds may be restricted to one or two levels. Similarly the healing from a cancer disease takes place simultaneously on the molecular, cellular and body levels; on the levels of the patient, his family and the work environment, and finally on the levels of immunology, chemotherapy or surgery. A good doctor reflects all the levels mentioned and many more, but his interventions will normally touch only a few of them at once.

The lower levels are characterized by a high degree of 'confirmation'. From this we can derive a first conclusion regarding quality and responsibility: Actions and perceptions concerning high-confirmation levels should be governed by the high ambitions of science and its traditional standards. Methods should be as perfectly "fitting keys" as possible. Fraud and incorrect handling of facts must not be tolerated. Just as physicians have to have a minimum training concerning the molecular level, agricultural administrators should have some testable, high quality proficiency on some 'lower' levels concerned. However, this methodological fitness and proficiency should not be mixed up with the *goal* of action. The goal remains, as indicated in the previous chapter, success and

By talking about goal-orientation we are invariably entering a *novelty-rich* domain. Contingencies, irreversibility, errors come in. The actor has to count with failure and destruction, but he also has the chance of innovation. Now the motto becomes "error-friendliness". Which is by no means identical with cowardice. The goal remains, as above, success and contagiousness.

Let us assume that to optimize contagiousness on the highest level of action a balance between confirmation and novelty according to Fig. 4.1 is to be aimed at. Then, to optimize both quality and responsibility, a balance of methodological fitness and error-friendliness[29] may be called for, (whereby the two complement each other, they do not outright contradict each other).

These ideas are not new. But by presenting them in a new language new approaches for testing them may be opened. This process of testing will have to

---

[29] Error-friendliness in technology is more than 'reliability' (which is normally rather associated with a certain inertia). As in biology such notions as resilience, barriers against damage-spreading, and mutations are important. To refer once again to M. S. Swaminathan (see note 25), contingency planning, risk distribution, disaster preparedness, and mid-season corrections were important factors in the Indian agricultural success (i.e. pp. 7–8).

establish in particular a more precise meaning of the terms involved. But since the terms qualify and responsibility are not well defined either (although we seem to use them without hesitation) one cannot realistically expect full precision.

References

Bush, Vannevar, et al. 1945: *Science—The Endless Frontier.* Reprinted by National Science Foundation (Washington, DC, 1960).
Cole, S.; Cole, J.R.; Simon, G.A. 1981:"Chance and Consensus in Peer-review", in: *Science*, 214: 885; cited in: Irvine, John; Martin, Ben R. 1983: "What Direction for Basic Scientific Research" (Manuscript, Science Policy Research Unit, University of Sussex, April 1983).
Khoo, T.C., 2005: "Water Resources Management in Singapore", Paper for the 2nd Asian Water Forum, Bali, Indonesia, 29 August–3 September 2005
Meyer-Abich, Klaus, M. 1979: "Die gesellschaftliche Wirklichkeit der Naturwissenschaft—Zum Problem der praktischen Wahrheit der Naturwissenschaft", in: Eisenbart, Constanze (Ed.): *Humanökologie und Frieden* (Stuttgart: Klett)
Paddock, W.; Paddock, P. 1967: *Famine 1975—America's decision: Who will Survive* (Boston: Little Brown).
Ramanujam, S., et al. (Eds.), 1980: *Science and Agriculture—M. S. Swaminathan and the Movement for Self-Reliance* (New Delhi: Associated Publ. Co.)
von Weizsäcker, Ernst, 1984: "Überraschung und Ordnung gibt Information", in: Schaefer, G. (Ed.): *Information und Ordnung* (Aulis: Deubner Verlag).

Chapter 5
New Frontiers in Technology Application: Integration of Emerging and Traditional Technologies

5.1 Preface

This volume is based on manuscripts, the majority of which were submitted to the ad hoc panel of specialists on "Integrated Application of Emerging and Traditional Technologies for Development". The panel was the first in a series of panels of the United Nations Advisory Committee on Science and Technology for Development and was held from 13–16 December 1982 at the *International Rice Research Institute* (IRRI), Los Banos, Laguna, The Philippines.[1]

The panel elected Dr. M. S. Swaminathan, Director-General of IRRI, as its Chairman. He steered the panel in the direction of collecting and designing pioneer projects to clarify and illustrate the innovative concept of blending new and old technologies. Dr. Aklilu Lemma, Senior Scientific Affairs Officer, United Nations Centre for Science and Technology for Development, served as Secretary of the Panel. He and Dr. Swaminathan, together with Dr. Clarence W. Bockhop, carried the main burden to make the panel an impressive success. Because of the significant contributions they have provided to the conceptual framework of the panel, the names of Drs. Swaminathan and Lemma are included as co-editors of the volume.

The Advisory Committee, in its third session from 1–8 February 1983, welcomed the Report of the Panel and considered it very important to disseminate and popularize it widely. The Committee also encouraged the Secretariat to publish a proceedings volume and to include a few more contributions solicited from WHO, FAO, UNIDO and perhaps others. The Advisory Committee also stressed the need

---

[1] Proceedings of the ad hoc panel of specialists of the *United Nations Advisory Committee on Science and Technology for Development* (ACSTD) on "Integrated Application of Emerging and Traditional Technologies for Development", held at the International Rice Research Institute, Los Banos, Laguna, The Philippines, 13–16 December 1982, in: von Weizsäcker, Ernst U.; Swaminathan, M.S.; Lemma, Aklilu, (Eds.), (Dublin: United Nations by Tycooly International Publishing Ltd.).

to concretize the theme by giving much attention to pioneering experience in the field.[2] As regards this proceedings volume, this was accomplished by substantially expanding (in Sects. 4 and 5 of the volume) Annex II of the Panel's Report, which included ten sketches of pioneer projects, both completed and projected ones. Yet it remains quite evident that the selection of projects showed a certain arbitrariness. The panel was very small and the possibility to solicit additional inputs without delaying the printing was rather limited.

In the meantime, ILO, UNIDO and UNCSTD have embarked on a joint project to prepare a systematic compendium or portfolio of successful and unsuccessful experiments and projects, for the purpose of reviewing past and present efforts and drawing lessons for future possibilities of blending new technologies with traditional ones with a view to raising the productivity of the latter. The Advisory Committee has also noted with satisfaction the indication of some Japanese organizations to organize, in 1984, follow-up action on the design, management and implementation of pioneer projects. Similar offers of collaboration and interest for further elaboration and implementation of the recommendations of the Los Banos Panel have also been made by the Chinese and Brazilian governments.

During the session of the Advisory Committee the strong desire was expressed to have the volume available for the next regular session of the Intergovernmental Committee for Science and Technology for Development, meeting from 6–17 June 1983. This made it unavoidable to request the authors to accept a very short deadline for the submission of their manuscripts, be they revised versions of the papers presented to the panel or even entirely new papers. I am impressed and deeply grateful that all authors have responded favourably to my request. In particular I wish to thank those authors who have prepared new papers at such short notice. I am also indebted to many authors who very generously agreed to some cuts or minor substantive changes.

The editorial work was greatly facilitated through the excellent work of Messrs Jörg Boltersdorf, David Eade and Paul Miller who did most of the linguistic and formal editing; Mr. Clemens Kaiser and Ms Cecilia Chan helped in establishing the subject index. Ms Sandra Schweighofer and Ms Joanne Bolanda took care of the voluminous correspondence and the re-typing of most manuscripts; and Christine Skrzek did the bulk of the pre-panel clerical work. I wish to thank them all.

Indebtedness is also expressed to Tycooly International Publishing Ltd., especially to Mr. F. O'Kelly for the excellent co-operation and the remarkably speedy printing and publishing of the volume in less than eight weeks. Finally, I wish to acknowledge the lively interest Professor Amilcar F. Ferrari, Executive Director of the United Nations Centre for Science and Technology for Development, has taken in the panel and in the preparation for this volume. His constant encouragement was felt throughout the work.

New York, 5 April 1983   Ernst U. von Weizsäcker

---

[2] For the full text of the Advisory Committee's report as far as it concerns the panel, see UN Document A/CN. 11/34 (4 March 1983), para. 35–43.

5.1 Preface

The first two panels of experts of the Advisory Committee on Science and Technology for Development have concluded their work. At least five more panels are being planned for the years to come. The Centre for Science and Technology for Development is proud to present the proceedings volume of the very first of this important series of seminars.

The common denominator is the strengthening of endogenous capacities for science and technology in developing countries. This, however, is an enormous task which must be approached from various angles. Countries need an endogenous policymaking structure for science and technology; they need an adequate scientific and technological infrastructure; they need endogenous capabilities for choice and acquisition of technology; and perhaps most important, they need the human resources to carry out the endogenous research and development. Furthermore, viable mechanisms and incentive structures for the financing of science and technology will be needed in the developing countries; a functioning, modern information base for technologies and scientific methods and results is needed; and the developing countries need to strengthen their domestic research and development and to link it in multiple ways with their productive sectors; last but not least the international co-operation among developing countries and between developing and developed countries needs to be strengthened, in accordance with the principle of collective self-reliance.

The eight needs which I have stressed above are in fact the key words for the eight programme areas of the operational plan for the implementation of the Vienna Programme of Action on Science and Technology for Development. This is also the legal frame for the work of the Advisory Committee including its panels of experts.

During the deliberations of the Advisory Committee at its second and third sessions, it became abundantly clear that the integrated application of emerging and traditional technologies provides an excellent chance for a type of innovation reaching out into the villages and rural areas of developing countries. We feel grateful that so many distinguished scholars and technologists from all around the world have devoted so much of their time to prepare their contributions to the panel and to this proceedings volume.

Chapter 6
The Environmental Dimension of Biotechnology

6.1 Eurosclerosis and Beyond

Europeans have started to fight Euro sclerosis.[1] 'Sleeves up, keep the Americans and Japanese from getting ahead'—that is the new message which you can hear from all sides.[2] The European Parliament devoted a week to the technological challenge.[3] The problem appears to be that, although people say we can run as fast as others, we have more disputes about the direction, and that environmentalists and other citizens express a lot of hesitation with regard to new technologies. There are good reasons for some hesitation and Japan and America also have disputes. Yet, their environmental concern has made them even more competitive. Let us have a closer look at the Japanese model (Shigetu Tsuru/Helmut Weidner 1985).

Leaving aside the important questions of culture and attitudes, Japan is good at key technologies and increasingly good at developing the scientific foundations of new technologies. Yet the Japanese environmental policy is considerably more stringent than ours.

The coincidence between technological modernity and stringent environmental standards is perhaps more than mere chance. At one time, most people, including Japanese industrialists, believed that there was a conflict between industrial progress and the environment. The Minamoto and other disasters forced Japanese industry to combat pollution at all costs. But the undesired shift towards pollution control turned into a blessing.

---

[1] This chapter was drafted and presented at a time when knowledge about the problems of agricultural genetic engineering was in its infancy. A mere ten years later, the alleged benefits e.g. of incorporating the BT toxin into crops had become very doubtful. At the time of republishing this piece in an anthology of the author's writings, i.e. in 2014, the thrust of an article written on this topic would have been completely different. E.v.W. 2014.

[2] This text was published in: Davies, Duncan, (Ed.), 1986: *Industrial Biotechnology in Europe* (Brussels: Centre for European Policy Studies): 35–45. The permission to republish this text was granted by Ms. Margarita Minkova on behalf of CEPS on 5 February 2014.

[3] European Parliament, Working Document A 2-109/85/B; European Parliament, Working Document A 2-110/85; European Parliament, Working Document A 2-108/85; European Parliament, Working Document A 2-89/85; European Parliament, Working Document A 2-106/85; European Parliament, Working Document A 2-99/85.

During the 15 years since the shift, technologies have emerged to decentralize production, to substitute energy and materials by information, and to substitute crude chemical processes by sophisticated biological processes. Rapid technological modernization has turned out to be the best strategy for meeting the stringent environmental standards. Or to put it differently: environmental policy was discovered to be a powerful whip for accelerating the technological progress. This may be the means whereby Europe can free itself from the lamented sclerosis.

6.2 Biotechnology: Environmental Pros

6.2.1 Agriculture

In agriculture, 'biologicals' have a very good sound among environmentalists. They usually denote integrated biological pest control measures, which are increasingly replacing poisonous and polluting chemicals. As in the Japanese example, it was growing environmental concern and the ban by the European Community of some pesticides that spurred the chemical industry into research on biologicals. About 15 % of the pest control research money in the industry of the *Federal Republic of Germany* (FRG) goes into biologicals, and their application is slowly spreading.

Further, the breeding of plants resistant to disease or to certain pests can lead to a considerable reduction in the use of poisonous chemicals. Josef St. Schell, Director of the Max Planck Institute for Plant Breeding Research (and head of the Genetics Laboratory of Ghent University) has developed—together with other teams around the world—a method which promises to transfer genetic material into higher plants. The prospects are good for using the tumour-inducing plasmid of *Agrobacterium tumefaciens* as a vector for genes coding for insecticides, fungicides, herbicides, high-value nutrients, or for resistance against microbial diseases (Schell 1985).

Easier than introducing the genes directly into plants may be to breed microbes to do the biochemical job. Monsanto, St. Louis, developed microbes containing toxins to kill pests and to let those microbes colonize the crop plants. Lidia Watrud and her colleagues were able to insert the gene coding for the well-known delta toxin of *Bacillus thuringiensis into Pseudomonas fluorescens,* which can colonize the roots of maize and other plants. Root-eating insect larvae ingesting the genetically engineered bacteria will die. Monsanto's proposal, submitted to the *Environmental Protection Agency* (EPA) for field-testing the system, is still under review.[4]

---

[4] Kolata, Gina, 1986: "How Safe are Engineered Organisms?", in: *Science*, 229: 34–35. (Brief article on the Philadelphia Conference on Engineered Organisms in the Environment: Scientific Issues, 10–13 June 1985).

In a similar fashion, science is attacking the problem of nitrogen fixation with good prospects. Scientists try to introduce into non-leguminose plants the ability to fix atmospheric nitrogen, through the symbiotic mycorrhiza bacteria. Since the process involves at least seventeen different bacterial genes, it seems much more plausible to introduce existing bacteria to associate with cereal crops. If this succeeds, nitrogen fertilizers could be supplanted, which would help to solve the nitrate pollution problem.

In the field of drought or salt tolerance, (Sondahl et al. 1985) Monsanto is said to have genetically engineered a drought-resistant maize strain and awaits permission to market it. It is evident that plants with pest tolerance, salt tolerance, reduced fertilizer demand, and better nutritional value at no additional cost are at first glance ecologically very desirable (Hauptti et al. 1985).

Different considerations are involved for herbicide-resistant crops. Robert Goodman of Calgene, in Davis, California, E. Jawashiat of Monsanto, and other teams are experimenting with inserting bacterial resistance factors against glyphosate herbicides—first into plant plasmids and then into tobacco, tomatoes, cotton, maize, soybeans, and other plants.[5] Inasmuch as this will make possible the application of herbicides which are effective in low doses (grams instead of kilograms per hectare), that research is welcome both from an economic and an environmental point of view. But if the herbicide in question is hazardous, such as atrazine, the target of research of Ciba-Geigy (USA), the fear arises that farmers may be induced to poison the soil and to inflict incalculable damage to the environment. If atrazine is banned in countries with high safety and environmental standards, the package of atrazine and atrazine-tolerant crops could still do a lot of damage in developing countries.

Biotechnology can greatly help to optimize the design of integrated farming systems, involving plants as well as animals and aiming at better resource recycling, through biogas production, fish ponds, the production of pharmaceuticals, intercropping, etc. The growing of 'energy crops' also has an environmental impact in that it slows down the depletion of non-renewable resources and alters the ecological structure of the countryside.

6.2.2 Industry

There are two areas for consultation: biological process technologies complementary to and partly even substituting for chemical technologies, and biological clean-up technologies. In the first area, the production of sophisticated biochemicals, including drugs by biological processes, replaces clumsier resource-consuming and

---

[5] Kolata, Gina, op. cit. See also Brill, W. 1985: "Safety Concerns and Genetic Engineering in Agriculture", in: *Science*, 227: 381–384, and the table of US R&D programmes for herbicide-resistant plants, in: *Genetic Technology News*, April 1984.

often polluting chemical technologies. Also bacterial leaching of metals from ores eventually becomes environmentally preferable to conventional methods. However, to date the environmental significance of these biological 'cleaner technologies' is considerably less than that of agricultural biotechnologies.

In the area of biological clean-up technologies, micro-organisms, including those genetically engineered, are being used for the detoxification of industrial wastes and for the routine purification of waste water. Hans Joachim Knackmuss of Wuppertal and K. Timmis of Geneva have succeeded in the laboratory in combining in one bacterium the genetic material for an entire chain of reactions that degrades poisonous aromatic compounds (Timmis et al. 1985). They were following the road opened by Ananda Chakrabarty, of General Electric in America, who took microbial samples from the poisoned surface of Love Canal and from other similar sites. He was able to isolate microbes which had adapted to the toxic environment. From these survivors, he extracted plasmids and inserted them into Pseudomonas bacteria. By exposing these to increasing levels of 2, 4, 5T, he bred bacteria able to degrade 90 % of the 2, 4, 5T in a contaminated soil sample in a laboratory test clean-up operation (Ghosal et al. 1985).

The necessary field tests for such organisms have not been permitted so far. There is a considerable debate among environmentalists whether one should deplore that general ban on the release of genetically manipulated organisms or if one should not at least permit exceptions for such cases as Chakrabarty's Pseudomonas bacteria. I shall return to that question later. However, even without the release of manipulated organisms there is considerable scope for biotechnology in clean-up and waste treatment operations. The poison-eaters can, of course, be used to detoxify dangerous waste as long as it is well contained.

Another environmentally desirable road of research is the breeding of anaerobic micro-organisms which could be used for special or general tasks in waste (water) treatment. Anaerobics would work with less energy input than aerobics, produce usable biogas as an output, and reduce the volume of sewage sludge left (Buchholz et al. 1985). But even without any genetic engineering environmental gains can be expected from gradual progress in biotechnology. Botho Böhnke of Aachen looked at the microbial composition of ordinary waste water tanks. He found that the bacterial population which is responsible for the cracking down of undesired chemicals could be greatly enhanced if there were no bacteria-eating bigger (eukaryotic) microbes around. So he designed a two-chamber system, with the first chamber hosting a hundredfold density of bacteria and the second one resembling in composition the ordinary tanks. With this two-chamber system (called adsorption-bio-aerated treatment or A–B-treatment), he was able to build considerably smaller and therefore cheaper water clarification plants which, most importantly, are extremely robust against changes in the chemical composition of the waste water (Böhnke 1982). Even shock bursts of toxic chemicals which would paralyze ordinary waste water treatment stations were easily absorbed: the high density of bacteria allows for a much greater variance of their biochemical arsenal for cracking and digesting alien chemicals.

Böhnke estimates that more than a hundred thousand positive genetic mutations will occur daily to respond to all kinds of chemical challenges. Moreover, the steady flow of fresh, living bacteria into the first chamber (which is another feature of the A–B-treatment) leads to a quick replenishment with bacteria even after pulses of highly toxic chemicals may have arrived. The second chamber receives largely detoxified water, and the final outflow may already have drinking water quality, depending largely on the content of heavy metals. A full-scale demonstration plant is already working in Krefeld and more plants are under construction.

6.3 Environmental Cons: Success is the Danger

Let me now turn to the environmental 'cons' of biotechnology. The construction and the release of genetically engineered new organisms require our serious concern. Since it is now clear that manipulated micro-organisms lack the vitality to proliferate to any significant degree in natural environments, safety regulations have been relaxed, and the public support for genetic engineering is now stronger than ever before. A different category of dangers now needs to be taken much more seriously. I am referring to the 'dangers of success'.

Small-scale agriculture has for thousands of years improved rather than damaged the environment of the European continent (Kramer 1982). It is only in the last 35 years that agriculture has become 'successful'. However, if you imagine a functioning Common Market with comparable incentives to introduce biotechnological routines into all farms, there may be very serious concerns with the loss of genetic diversity and environmental degradation in general.

If herbicide-resistant crops are cultivated on a large scale, and cheap herbicides are available, then it is possible that very large areas may be turned into botanically dead soil on which only the man-made, herbicide-resistant crops will grow. If the herbicides are persistent, this situation may persist for years. A related hypothetical problem arises if plant species are being successfully bred which are vitally stronger than their natural competitors. Until now it was clear that man-bred crops have no chance in natural habitats. But if they are designed to be salt-tolerant, pest-tolerant, or fertilizer-independent, the situation may change. It should be emphasized, however, that vitality in natural habitats requires much more than single features like salt-tolerance; especially the complex factors governing fertility and the potential for geographical expansion are unlikely to be optimized in cultivated crops, since they do not relate to man's breeding objectives. Even integrated biological pest control could conceivably become an environmental problem. If successful against all enemies defined as such at the time of its application, it could later become apparent that positive functions of those 'enemies' had been overlooked, and then it could be too late.

Another 'danger of scale' deals with plasmids, which become abundant in genetic engineering laboratories, in industrial fermenters, in waste water treatment plants, and on farms. It has to be assumed that many of those plasmids will carry survival factors or vitality factors for various environmental challenges; given a high density of such factors, we might in theory be faced again with the fears dealt with at Asilomar, i.e., the fears of a doomsday bug or at least of a toxic bug proliferating in the natural environment.

6.3.1 Evolutionary Concerns

Neo-Darwinism relies on two well-known factors—mutation and selection—and one almost forgotten factor—isolation, an indispensable driving mechanism for the origin of species. If the fittest species were able to expand unchallenged, not many species would remain. One primitive organism could have prevented the rest of the world from developing if it were not somehow restricted. The complexity of ecosystems and the variety of species rely heavily on the existence of barriers of different kinds, but also on the preservation in the gene pool of 'less fit' alleles. Isolation stands for the necessary coexistence and 'co-evolution' of the fittest with the less fit. The 'survival of the less fit' is just as important for Neo-Darwinism as the 'survival of the fittest'.

Man has fought an increasingly successful war against isolation. In agriculture, field sizes were enlarged. Eucalyptus, potatoes and cereals, once localized plants, were spread out over the world and became dominant for many artificial ecosystems. Germs are transported across continents. An unprecedented wave of simplification and standardization of ecosystems is driving thousands of plant and animal species into extinction and is causing a huge loss of genetic diversity even within the remaining species. It is almost certain that the potential for further evolution of the biosphere and its resilience against climatic and other shocks will suffer from these massive losses.[6] Biotechnology by its very nature can reinforce this trend because of its need to be useful and effective in as many ecological conditions as possible.[7]

6.3.2 Some Thoughts on Regulation

It seems evident that some sort of regulation will be necessary for the wide range of biotechnologies which will be applied in our future world—certainly no less

---

[6] For a balanced discussion of this much debated point, see Pimm, Stuart L. 1984: "The Complexity and Stability of Ecosystems", in: *Nature*, 307: 321–326.

[7] von Weizsäcker, Ernst, "Konsequenzen der Gentechnologie aus der Sicht moderner Evolutionstheorie", written and oral evidence to the Commission of Inquiry of the Federal Republic of Germany Bundestag on "Prospects and Risks of Genetic Engineering" ("Chancen und Risiken der Gentechnologie").

than the introduction of the automobile has required. In the case of the automobile, virtually no year in its hundred years' history has elapsed without new regulations. This will no doubt be the case also for biotechnology.

In the US, a Coordinated Framework for Regulation of Biotechnology has been prepared by the Office of Science and Technology Policy in the Executive Office of the President of the United States.[8] That framework consists primarily of a matrix with seven major headings:

1. licensing and other pre-marketing requirements,
2. post-marketing requirements,
3. export controls,
4. research and information gathering,
5. patents,
6. air and water emission standards,
7. requirements for Federal Agencies.

Besides the matrix, there are policy statements of the three federal agencies involved (the *Environmental Protection Agency* (EPA); the *Food and Drug Administration* (FDA); and the *Department of Agriculture* (USDA), and a Scientific Advisory Mechanism to substitute for the Recombinant DNA Advisory Committee that was established by the National Institutes of Health in 1974.)

In substantive terms, the proposed Framework foresees a considerable relaxation for the experimental phases, including field- testing of biotechnological innovations. However, 'premanufacture notices' (PMN) are required for all new chemical substances and, by implication, new organisms. The EPA must assess the potential risk to be regulated.

The public discussion of the proposed Framework is still going on,[9] but the critical comments received do not touch its substantive core, so that no drastic changes are expected to be made. (Meanwhile, the Occupational Safety and Health Administration has also produced guidelines on occupational safety and health in the field of biotechnology. They fit into the framework.) There is a likelihood, however, that in the institutional power struggle, the FDA may win over the EPA. But the concerns which I have expressed in the two preceding sections are not really met by the US proposals. In fact, any regulation that sees the risks only in immediate hazards and does not specifically address the problems of large-scale uses is likely to be deficient.

In Europe, there are various national regulations in existence, on drugs, on genetic engineering research, on the marketing of products, etc. But there is no specific environmental regulation, except for the general restriction against release into the natural environment of any genetically engineered organisms. On the

---

[8] Proposal for a Coordinated Framework for Regulation of Biotechnology, *Office of Science and Technology Policy Notice*, 31 Dec 1984.

[9] An important response was: Covello, Vincent T.; Fiksel, Joseph R. (Eds.), 1985: *The Suitability and Applicability of Risk Assessment Methods for Environmental Applications of Biotechnology* (Washington, DC: National Science Foundation).

international level, there is, of course, the *Committee on Genetic Engineering* (COGENE), originated by the *International Council of Scientific Unions* (ICSU), but it has no legislative powers. Fortunately, the OECD's Ad Hoc group on biotechnology safety came to a preliminary agreement on some substantive issues at its meeting in Paris 2–5 December 1985.

At the EC level there is the Biotechnology Action Programme adopted by the Council in March 1985, which lists regulation as one out of six priorities. However, it does not cover the concerns expressed in this chapter. Perhaps more pertinent is the new general Directive on environmental impact assessments. But by its very nature, and owing to the fact that it will take years before any practical experience with its implementation can be obtained on national levels, this Directive can only be a corollary to biotechnological regulation.

It is time, so it seems, for the European Community to take the initiative on European regulation.[10] In doing so the EC ought not to make highly specific restrictions and bureaucratic regulations, but develop a common understanding of the real problems involved and of the lacunae in our knowledge. Regulation ought not to eliminate all risk, but to make sure that the risks that remain are of the magnitude of other risks in ordinary life. This would be equivalent to acknowledging the evolution principle of 'isolation'.[11]

In this state of affairs, where there is an obvious demand for international agreements on regulation, but where much basic knowledge is yet to be developed, I could imagine that the European Community might set up a team of experts on the questions of regulation in biotechnology. The team could use the valuable work of COGENE, the OECD Ad Hoc group, earlier WHO work, and national experience, and could develop concrete proposals to fill the vague frame of the European Biotechnology Action Programme.

Since the main issues are of great public importance, the impression ought to be avoided that the important regulation questions are being handled behind closed doors. Hence, a more public event will also be welcomed, such as the European Conference on Biotechnology Regulation. Industrialists in democratic societies need not be afraid of the public nature of the debate. After all, automobile regulations are not developed in the dark either. Rather, the active involvement of the EC may be taken as a symbol of the new spirit of Europeans to face modern challenges and to overcome the so-called Euro sclerosis.

---

[10] See a draft paper: "Community Regulations Impinging on Biotechnology", BSC/4/2.4.2, March 1985, available from the Concertation Unit for Biotechnology in Europe (Brussels: Commission of the EC, 1985).

[11] This refers to the step-by-step, cumulative 'learning' approach, c.f. Clark, William C. 1980: *Witches, Flood and Wonder Drugs. Historical Perspectives on Risk Management* (Vancouver: University of British Columbia, Institute of Resource Ecology).

References

Botho, Böhnke, 1982: "Mit Bakterien-Kolonien gegen verschmutzte Abwässer", RWTH-Themen (Aachen: Technische Universität), No. 2: 10–12.

K. Buchholz et al., 1985: "Technischer Fortschritt und fortschrittliche Technik: Wie nützlich ist die Gentechnologie?", Materialien der Studiengruppe "Gesellschaftliche Folgen neuer Biotechniken" der Vereinigung Deutscher Wissenschaftler: Mimeo: 19–27.

D. Ghosal, I.-S. You, D.K. Chatterjee, A.M. Chakrabarty, 1985: "Microbial Degradation of Halogenated Compounds", in: *Science*, 228: 135–142.

H. Hauptti et al., 1985: "Genetically Engineered Plants: Environmental Issues", in: *Bio/Technology*, 3: 437–442.

Peter, Kramer, 1982: "Biotische Viefalt: Ihre Evolution, ökologische Steuerung und Beeinflussung durch den Menschen", Habil.Schrift, Universität Essen.

Josef St. Schell: "Leben mit fremden Genen", in: Naturwiss. Rundschau, 36: 254–260.

Maro, Sondahl; William, Sharp; David, Evans, 1985: "Biotechnology for Agriculture of Third World Countries", in: ATAS Bulletin, Tissue Culture Technology and Development (New York: United Nations).

Shigetu Tsuru/Helmut Weidner, 1985: *Ein Modell für uns: Die Erfolge der japanischen Umweltpolitik* (Köln: Kiepenheuer & Witsch).

K.N. Timmis et al., 1985: "Analysis and Manipulation of Plasmid-Encoded Pathways for the Catabolism of Aromatic Compounds by Soil Bacteria", in: Donald R. Helsinki et al. (eds.): *Plasmids in Bacteria* (Pleumin: 719–739).

Chapter 7
Not a Miracle Solution but Steps Towards an Ecological Reform of the Common Agricultural Policy (CAP)

7.1 Introduction

The Copenhagen Summit demonstrated how urgently the EEC needs a solution to the CAP crisis. The CAP crisis is certainly more than a crisis of EEC finances. Solutions have to go much beyond fiscal changes or agreements on national shares of the burden. The very structure of a system inducing overproduction, environmental degradation and huge financial burdens has to be altered without jeopardizing the potential for progress and innovation.[1]

In conventional economics 'progress' means more production. But what is progress in a situation of overproduction? Conventional economics would demand the closing down of the least productive units until the market equilibrium at world market prices is reached. That would leave the efficient producers the necessary room for further progress in productivity. But what if production in less productive areas and by economically sub-optimal methods is a cultural or sanitary good or an ecological necessity? Conventional economics, stretching its dogma only a trifle, would suggest to prohibit environmentally destructive techniques and to let the state or some other public interest agent pay for the public goods.

But what if the producers, the farmers, insist that their entrepreneurial pride cannot accept income from 'bureaucratic' sources depending on political good will and changing majorities? Well, there may be no need to inform them about the bureaucracy of the present CAP and about the political struggles every year to fix the internal 'market' prices; or to inform them about other economic sectors like education, culture, military or transport depending considerably more on state money. Besides, the economic situation of many farmers has become so desperate that they have to accept any life belt (Oddly, it is not the minority of cereal farmers that are the most troubled ones, but they are the ones who seem to have had the strongest influence on the German government when it was blocking further progress at the Agriculture Council and the Summit).

---

[1] This essay was published in 1988 by the Institute for European Environmental Policy, Bonn-London-Paris. It resulted from a study commissioned by the Federal Ministry for Environment, Nature Protection and Nuclear Safety, Bonn, which is in the public domain.

Agreement seems to emerge that agricultural production is not an economic activity alone. Following the above reasonings we have to think of solutions to the CAP crisis which involve some sort of direct payments to farmers for 'producing'— or maintaining—other non-food goods including the protection of the environment. A closer look at this idea of direct payments shows, however, that taken alone it cannot be a quantitatively satisfactory answer to the crisis. Incomes much below present levels would be a disaster for millions of farmers in Europe. But taxpayers are hardly prepared to pay the necessary thousands of millions of ECU[2] annually— as they do for the present CAP—for just another CAP even if the new CAP is less wasteful and ecologically more acceptable. Taxpayers and states want to get rid of the huge burden as soon as possible.

Therefore one has to look at mixed solutions in which consumers pay for better food, the society pays for a better environment and farmers develop additional income opportunities.

In this essay, a policy mix is proposed with a certain selective preference for elements that promise to be good for the rural environment. The ambition is not to provide a miracle solution to the immense problem of the CAP reform. But it is hoped that each of the elements listed in the next chapter makes sense in the context of the CAP reform. And a preliminary semi-quantitative assessment of the sum of all sixteen elements seems to indicate that taken as a package these elements would be acceptable for farmers, good for the taxpayer and excellent for the rural environment.

The essay has many fathers. David Baldock of IEEP's London Office has guided the Institute into the field of the CAP and the environment. Hermann Priebe has taught the author about the not so obvious and in part ironic ways the CAP determines farmers' incomes under different conditions (Priebe 1985; Priebe/von Weizsäcker 1987). Thieny Lavoux of IEEP's Paris Office conducted major pertinent studies and contributed the necessary insights about the French perception of the CAP and the rural environment. Francois Roelants du Vivier who was Rapporteur on agriculture and the environment at the European Parliament has proven that a policy mix with a strong inclination towards organic farming can find near unanimous support from all political sides (European Parliament Resolution 'Agriculture and Environment' of 19 February 1986; Roelants du Vivier 1987). Wolfgang Haber, main author of the monumental report on agriculture and environment of the German Council of Environmental Experts (SRU 1985) has spent many valuable hours discussing with the author,—as have uncounted other persons, farmers, environmentalists, politicians and others. The responsibility for all mistakes remains, however, with the author. A German version of the essay was written in January 1987 as a contribution to the Festschrift to honour Hermann Priebe (von Urff/von Meyer 1987); this English version was, however, entirely rewritten.

---

[2] ECU (European Currency Unit) was the precursor to the Euro, worth roughly 2 deutschmarks.

7.2 Sixteen Elements of an Ecological Reform of the CAP

This chapter outlines 16 elements of an ecological reform of the CAP. About each one of these elements one could easily write a book. By allowing just one page per element this essay inevitably has to simplify if not caricature matters.

7.2.1 Food Quality and Consumer Information

European consumers show an increasing consciousness for food quality. Many complain about watery meat, tasteless eggs, chemical residues on fruits, nitrates in drinking water and allergies caused by food. Such concerns are not always justified. In fact, nutrition in Europe can be assumed to be richer than ever before. Nevertheless the new consciousness can be used as a driving force for agricultural reforms in the desired direction.

Most consumers want to know about additives and chemical residues in their food. Some consumers want even more information, e.g. about the conditions under which animals are raised or about ecological characteristics of the farm. And why should not they? Consumers seem prepared to pay higher prices for food they judge to be better for their health. Some are even prepared to pay a price differential just for the sake of animal welfare or of a healthy environment. 'Organic' wheat can fetch three times the EEC guarantee price. Although this example cannot be generalized for many other farm products, and the quantitative relation in the example is unlikely to remain when the supply of organic food grows much further, the opportunities seem to be considerably larger than most of us believed just a few years ago. It seems noteworthy that the 'organic' market tends to channel the consumer's money more directly and to a higher extent into the farmer's pocket than the EEC market regulation mechanism.

The best instrument to achieve a bigger market share of high quality or 'organic' food is rather not any regulatory action against ordinary food—which would be unjustified—but consumer information and product labelling. Indeed there are many successful examples of required information about food additives. Untreated lemons have become a product category as have eggs from freely roaming hens. A European label for 'organic' products and the introduction of standards visible on food products for environmentally desirable production methods is one of the proposals made by the European Parliament in its quoted Resolution of February 1986. Obviously one should not overestimate the quantitative effects in the near future of such proposals. But as the consumer is meant to be the master of the game called market economy one should start with him when listing elements contributing to a CAP reform.

7.2.2 Limiting the Use of Certain Chemicals

The use of certain chemicals can be prohibited or quantitatively limited to protect the consumer and the environment as has been done several times in the past. Directive 79/117/EEC prohibits the use of certain pesticides, and more recently the use of hormones in animal nutrition was banned. The drinking water Directive 80/778/EEC sets upper limits for nitrates and pesticides in drinking water which should lead to restrictions in their use in water protection zones and possibly beyond. Even if one does not agree with all details of this Directive, the principle remains undisputed that chemicals have to be controlled not only in food but, perhaps more importantly, in drinking water.

To limit the use of chemicals may lead to a decrease in food production, at least temporarily. Even if this appears desirable in the present overproduction crisis of the CAP one has to be aware of the possibility that this eventually leads to rising food prices.

Besides regulatory actions on chemicals there is also the possibility to reduce their use by individual counselling of farmers. During the 'good years' in the sixties and seventies many farmers became accustomed to using very high quantities of both pesticides and fertilizers. By suitable crop rotation and by determining for each individual field the optimum level of the application of chemicals, the use of these chemicals can be drastically reduced. This concept, promoted as integrated crop protection even by the chemical industry, constitutes certainly an important improvement over the present situation. (For a comprehensive German collection of environmental legislation in agriculture, see Hötzel 1986).

Intensive livestock breeding, often referred to as 'animal factories', arouses concern by consumers, animal rights groups and environmentalists.

To meet the consumer's concerns about food quality it would be sufficient to introduce and enforce a new label for meat or other animal products from intensive livestock. This may or may not lead to a price differential in favour of lower intensity husbandry.

Animal rights groups are plainly against 'inhumane' animal rearing. In Switzerland they have succeeded in a ban of specified mass breeding techniques. In Germany they tried, unsuccessfully so far, to do the same via an amendment to the Federal Animal Protection Act. It can be expected that in more countries such movements will gain influence on the intensive livestock debate. But recent Swiss innovations in 'humane' stables seem to indicate that the productivity does not necessarily suffer. In the context of the CAP reform one should therefore not expect a major quantitative contribution from this side.

7.2.3 Limiting Intensive Livestock

More significant in quantitative terms are the environmental aspects of the problem. Manure disposal problems, ensuing water pollution, air pollution (methane) and nuisances from odours figure most prominently in this debate. In the Netherlands a general freeze against further intensification was introduced one night in November 1984, after mounting concerns with the disposal of liquid manure and relating problems. Now livestock farmers have to prove that they can dispose of the manure safely be it on their own grounds or on land 'hired' for this purpose. Two German Länder, Lower Saxony and Northrhine Westfalia have imposed manure disposal restrictions which are seasonally differentiated. Also the *polluter pays principle* (ppp) seems likely to be introduced in intensive livestock farms in the Netherlands and other EEC countries. Some people feel the solution rather lies in import restrictions for protein rich animal fodder, chiefly soybeans, but it is difficult to imagine how such measures can be legally defended.

To summarize this element: The public acceptance of intensive livestock farming in its present form is dwindling. Environmental considerations will by necessity lead to further restrictions which are likely to affect the productivity. But as pork and poultry do not fall under the FEOGA price guarantee system, the quantitative effect of reducing the financial burden of the CAP will be limited.

7.2.4 Nitrates

In the context of implementing the Drinking Water Directive, Community countries were faced with the problem that wells and rivers in modern farming areas had nitrates concentrations well above the recommended 25 mg/L and often above even the limit value of 50 mg. Mixing waters from different sources to comply with the Directive is obviously a rather unsatisfactory strategy. Besides, drinking water pollution is not the only problem caused by nitrates. Overfertilization is also a problem for grassland habitats (low-nutrient grassland supports a much greater variety of species than uniformly fertilized meadows) and for freshwater eutrophication.

It seems obvious that some state interventions are unavoidable to bring nitrate levels, both from commercial fertilizers and from manure, down to healthy levels. A simple way would be a general levy on commercial fertilizers, as proposed e.g. by the German Council of Environmental Experts, SRU; but that measure would not discriminate between appropriate fertilization and overfertilization, it could hit small farmers more than large ones and it would have no effect on animal manure. A measure limiting fertilizer application regardless of its origin and according to locally variable ecological criteria would be ideal but seems difficult to establish and to put into practice. (An original suggestion came from the Rainbow group in the European Parliament: to ban or to limit the use of stalk-shorteners, a measure

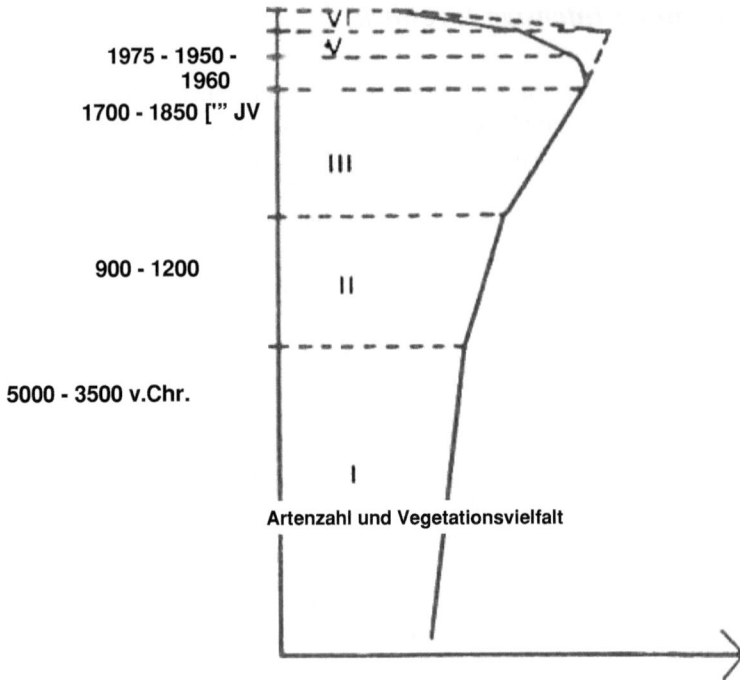

Fig. 7.1 Schematic representation of plant species and societies over the last 7000 years. *Dotted line* (after 1700) indicates enrichment through neophytes (imported plant species). From Sukopp/Hampicke (1985). This figure is adapted from Fukarek, F. 1980: "Über die Gefährdung der Flora der Nordbezirke der DDR", in: *Phytocoenologia*, 7: 174–182

which would make overfertilized cereal crops vulnerable to heavy rain and other usual weather stresses. But there is hardly a juridical basis conceivable for this astute proposal). Hence the present political discussion is essentially limited to nitrate restrictions in water protection zones. In some countries farmers are entitled to a compensation for their enforced restraint in fertilizer use (see Sect. 7.2.6).

7.2.5 Applying and Enforcing Existing Environmental Legislation

In the past farmers were considered to be the guardians of the rural environment. To a large extent this was indeed justified. Figure 7.1 shows how the plant species variety is believed to have increased in 5000 years, chiefly owing to habitat enrichment by small size farming.

The upper end of the figure, however, shows how modern farming (and other factors of modern civilization) have inflicted unprecedented damages to species variety. In this new situation it is clearly no longer justified to give agriculture any exemption from the application of existing environmental policy principles, notably the polluter pays principle, and of concrete legislation such as water protection laws, environmental impact assessment (Directive 85/337/EEC) and habitat protection in the context of the Birds Directive (79/409/EEC). There should be no CAP reform which would allow continued negligence about existing environmental legislation.

It can be assumed that a comprehensive and strict application in the agricultural sector of existing environmental legislation would lead to drastic changes in the economic conditions of production. Productivity may overall go down—although organic farming experience seems to show that productivity (per hectare—not necessarily labour productivity) can be kept at very high levels.

7.2.6 Compensation Payments

Elements 2–5 amount to considerable restrictions severely affecting the farmers' potential income. This is why farmers' representatives invariably raise objections against applying existing and creating additional environmental regulations for the agricultural sector. In the Southern countries of the Community, in Ireland and in South Germany farms typically operate under enormous economic difficulties. Many thousands of marginal farmers would have to give up if further squeezed by environmental restrictions. This would be tragic because their overall environmental effect tends to be positive. Hence it is absolutely understandable that one thinks of compensation payments for restrictions affecting the productivity of farmers.

The German Water Management Act foresees in paragraph 19(4) that farmers observing restrictions to protect water resources may be entitled to an adequate compensation. More concretely, the German Land Baden-Württemberg is introducing the controversial 'water penny' in January 1988 by which water consumers are charged a levy (approximately 10 Pfennig/m^3) which will be used to pay farmers in water protection zones a compensation of around DM 300 per hectare and year for the reduction of nitrates to an extent that the Drinking Water Directive values will be met.

Compensation arrangements of this type, however, have serious shortcomings. They are legally rather problematic in that they depart from the polluter pays principle which in turn is rooted in our fundamental beliefs in justice. It seems unlikely that compensation payments will survive many years of legal controversies originating from those who have to pay in the place of the polluters.

Another shortcoming is that poorer EEC countries are hardly in a position to pay 150 ECU per hectare and year to all small farmers suffering losses from environmental restrictions; and a Community instrument of compensation payments is difficult to imagine (see, however, next sub-chapter).

It should be noted that compensation payments per se are not an element of an ecological reform of the CAP, but they may in some countries be an indispensable condition for political progress.

7.2.7 Extensification Schemes

In the early seventies, the EEC realized that low intensity farming may have a merit of its own. At that time it was rather concerns with the depopulation of mountainous and other less favoured areas that led the Community to adopt the Less Favoured Areas Directive (75/268/EEC) (in German customarily referred to as Bergbauernprogramm). This programme, however, had mixed results as it seems to have encouraged overgrazing in Britain and Germany, while it certainly has contributed to sustain some desirable traditional farming systems, e.g. in France.

In 1985 new thoughts were introduced by Article 19 of a new Agricultural Structures Regulation, but it took two more years of negotiation until in March 1987 a second Regulation (1760/87 EEC) was agreed, allowing FEOGA contributions, up to 100 ECU per hectare and year. They are meant for schemes supporting environmentally desirable farming practices, not necessarily extensification, in *Environmentally Sensitive Areas* (ESA). Most countries have yet to designate ESA's and it waits to be seen how well the new programme will be accepted by the farming community.

However, there are remarkable for-running experiences in several countries which indicate that at least in less favoured areas farmers are quite susceptible to the new opportunity—for lack of other income prospects. One of the oldest programmes of this type was introduced in Bavaria under the name of 'meadow breeder programme' which offers around DM 300 per hectare and year for farmers maintaining their grassland and not cutting the grass between end March and end June so that certain rare and delicate birds (such as curlews, redshanks, snipes) are not disturbed during their breeding period. Another programme initiated in North Rhine Westfalia by the University of Bonn gives farmers 10–15 Pfennig/m^2 (equivalent of DM 1000–1500/ha) annually for not applying any chemicals on margin strips of their cereals fields to help guarantee the survival of wild plants and animals that accompanied traditional agriculture in Central Europe. Also in Britain there is a successful history of 'management agreements' of similar types. David Baldock of IEEP has collected the management agreement experience from four EEC countries (Baldock 1986).

A rather unexplored but promising type of management agreement relates to crop rotation. Rotation schemes differ with regard to soil conservation, chemicals demand, leaching rate and other environmental criteria. Farmers often choose sequences with rather small economic advantages but major ecological disadvantages. With rather small incentives they could therefore be induced to opt for the environmentally preferable crop rotation (Priebe/von Weizsäcker 1987).

7.2 Sixteen Elements of an Ecological Reform of the CAP   75

Extensification schemes have several advantages: They are a quick, specific and reasonably cost-effective instrument of nature conservation, they help cutting surpluses and they avoid major income losses of farmers. But there are also shortcomings. In some places the bird populations did not stabilize after the introduction of the habitat protection schemes; and there are so far hardly any extensification schemes for high intensity agricultural areas where a habitat recovery programme may be most urgently needed. Finally the basic philosophy of extensification agreements is somewhat problematic: the schemes can result in financial rewards for inaction. Juridical objections similar to those mentioned for compensation payments can theoretically be raised.

7.2.8 Payments for Ecological Work

Nature protection is not only a matter of specific inaction or restraint. Positive work can equally be done and honoured, such as the cultivation of hedges, the creation of ponds, the rearing of rare animals or the warding of nature protection zones. Management agreements can also encompass such positive tasks. Obviously no juridical objections would be raised against payments for positive ecological works.

Instead of paying for the work towards some ecological objective one may as well think of paying a premium for achieving and maintaining the objective. One could, e.g. pay a premium of, say, ECU 5000 annually for each successfully breeding pair of storks or curlews, both rare and ecologically very demanding birds: where they survive the entire local ecosystem can be considered as healthy. Another idea, already explored in the British Peak District, is to pay an incentive premium for species variety. The number of species of flowers, birds or butterflies can rather easily be determined, and cheating is almost impossible. A major advantage lies in achievement premiums as opposed to just work payments: farmers themselves would have a strong incentive to learn about the ecological properties—and vulnerability—of their land. The disadvantage is obvious: a lot of luck is involved especially with migratory birds, so the premiums contain an element of lottery. Another disadvantage is the cost of monitoring.

There is no automatic link between ecological work and the reduction of agricultural surpluses. And it is hardly conceivable that European societies are willing to spend amounts for nature protection which would compare with the present size of FEOGA. Nevertheless, for a certain number of farmers, a second income from nature protection may be the only realistic way of economic survival.

7.2.9 Set-Aside

Set-aside—the taking land off agricultural production—has become a favourite idea of agricultural experts concerned with the CAP reform. As said in the introductory chapter, set-aside conforms best with conventional economics. From the fiscal point of view it is attractive because it does not imply recurrent payments. And for the wilderness lobby it looks like the only attractive solution anyway. IEEP has—together with the Council for the Protection of Rural England, CPRE—organized a European conference on the topic, the papers of which have been published (Baldock/Conder 1988).

However, from the environmental point of view set-aside is of limited value only, and a general CAP reform strategy predominantly based on it could even be negative. The economic logic of set-aside would demand taking marginal farmland out of production as fast as possible and to rather increase the farming intensity in the remaining areas. Marginal agricultural land turned into forests or just uncultivated land is not automatically an ecological gain. In Mediterranean countries the likelihood of forest fires would be enhanced due to lack of people living in the area and helping to control the fires. Similarly soil erosion prevention and preventive actions against snow avalanches are rendered more costly and more difficult if the respective mountainous areas lose their agricultural backbone. This is why the Swiss are so heavily subsidising their hill farmers. In Germany the landscape and its ecological characteristics would become more uniform. For Britain, with her many deforested hills the ecological effect of set-aside-based afforestation seems much more positive. To generalize this observation: when new trees fill forest openings (as it would be the case in German hill areas) the ecological effect would be predominantly negative; when new forests grow on otherwise barren hills, the effect is likely to be positive. It is rather the length of forest edges—and not the total area of forests—that characterizes an ecologically healthy and species-rich landscape. Hence ecologically motivated set-aside premiums might be conceived which reward the creation of forest edges. Similarly where the abandonment of agricultural land would help recreating valuable wetlands, it could be encouraged while set-aside leading to less environmentally sensitive irrigation should rather be avoided, certainly not rewarded.

The German council of experts on environmental questions, SRU, in its monumental study on environmental problems of agriculture proposed an additional strategy for the recovery of wildliving species (SRU 1985: 309). It proposed the creation of a nation-wide network of ecological corridors which would cover at least ten per cent of the total area in each part of the country, including high intensity agricultural zones. Although the proposal was not very explicit about the inclusion in the calculation of existing unpaved roads, river banks, etc., it clearly would require considerable set-aside in high productivity areas. A wonderful idea but it would imply very high compensation payments or a rather problematic policy of expropriation or of management restrictions by the state.

To summarize: The environmental assessment of set-aside strategies is no simple affair. Neither a full steam set-aside policy nor a general refusal of set-aside is a reasonable attitude in the context of an ecological reform of the CAP.

Concerning the economic and social aspects (which are not the main focus of this essay) set-aside strategies may be a mixed blessing, too. Social and regional discrepancies would be rather enhanced than diminished, and the ultimate economic effect is far from being certain. A further concentration of agriculture in high productivity areas is likely to accelerate the industrialisation of the whole sector which could under the current biotechnology revolution lead to even higher surpluses (Hagedorn et al. 1986).

7.2.10 Prepension Scheme

Closely related to set-aside are prepension schemes for aging farmers to induce them to end their production for the market. Germany has successfully promoted this idea at the EEC level. Although in some cases this is the only solution available for older people, it should rather not be regarded as an element of an ecological reform of the CAP.

In Mediterranean countries and to a small degree in South Germany prepension schemes might, however, help accelerate the succession of generations on the farm rather than the abandonment of the farm. In that case benefits for the economic viability of the region and for the protection of the environment can be expected. On the other hand, for the same reason, one should in such cases not expect any reduction in agricultural production. Rather the opposite, owing to the propensity for modernization in the younger generation.

7.2.11 Biotechnology

Agricultural surpluses can also be reduced by converting agricultural production into the production of industrial raw materials. This idea has caught the imagination of the European Biotechnology Community and also of farmers looking for new business opportunities (Rexen/Munck 1984). First calculations have, however, led to the insight that bioalcohol in Europe may be an even more costly adventure than the present cereals market. The situation is not much better for other bulk chemicals for industry.

Besides, there are major environmental objections against large-scale agricultural production of industrial raw materials (IEEP—Memorandum 1987):

1. As long as the products of the land are consumed by humans or animals, there are a number of in-built constraints on the use of agrochemicals and Fertilizers beyond certain limits. Once products are turned into industrial feedstock, such restrictions could theoretically be relaxed or, in some cases, given up—to the detriment of the soil and water quality and with potentially adverse effects on fauna and flora.
2. Industrial clients of agriculture have a general tendency to demand large quantities of products all meeting the same standards. This could in theory reinforce the prevailing trends towards large field sizes, further mechanization, excessive reliance on monocultures and heavy use of pesticides. Such trends are all undesirable from the environmental point of view.

3. Most of the crops potentially suitable for industrial purposes are those grown intensively on arable farms with higher than average environmental damages; these crops include maize, sugar beet, rape seed, and cereals. To artificially maintain or substantially increase the demand for such crops cannot be considered desirable from the environmental point of view.

There are, however, also some positive aspects and opportunities:

1. As said in the set-aside context, afforestation is a good thing in many parts of Europe. If such afforestation is considered as a renewable raw materials strategy and if it is done under strict ecological criteria it certainly remains a good thing. For semi-arid areas it would also have side benefits in water retention, micro-climatic improvement and tourism. Its economics are quite tricky and cannot be discussed in this essay.
2. Some specialty chemicals for pharmaceutical, cosmetic, fibres and dye productions may be an interesting opportunity for some small farmers as is already the case for certain spices. Typically, such specialty chemicals require rather specific climatic conditions often found in "less favoured areas".
3. The ideas proposed by Rexen and Munck and their followers in the EEC should at least be explored experimentally under different climatic and economic conditions. It is well possible e.g. that their idea of small-scale rural bioindustries may prove viable under certain conditions.

The field of biotechnology is much broader than the renewable raw materials idea. In the context of an ecological reform of the CAP also the scope for biological pest control, development of pest-resistant crops, biotechnological processing of animal manure and detoxification of pesticide residue should be explored. At first sight the economic and ecological impacts of these improvements look small for the time being but could be large in the future.

The deliberate release into the environment, i.e. the agricultural use of genetically engineered organisms is a rather controversial issue. The Institute for European Environmental Policy has taken a rather cautious attitude (Lavoux 1987; IEEP—Memorandum 1987; von Weizsäcker 1988). In the context of this essay it would lead too far to go into any details.

7.2.12 Part-Time Agriculture

The CAP was conceived for full-time agriculture. The Mansholt Plan was clearly designed for high-efficiency farmers concentrating all their attention on the maximization of food production. The socio-economic goal was to keep the farmer's income always in the vicinity of comparable elevated industrial wages. Part of the crisis consists of the fact that this objective has become illusionary for more and more farmers. As long as they had the option to leave the land and accept some employment in the secondary or tertiary sectors, the situation seemed

7.2 Sixteen Elements of an Ecological Reform of the CAP

manageable. Under the present unemployment rates with many young adults being much better qualified than aging farmers for the rarer jobs in the cities, the situation has become rather desperate for farmers with dwindling income.

In reality, however, the situation is far less dramatic owing to the fact which is not so widely known, that in many European countries part-time agriculture has gained ground. According to Priebe (1985: 177), only one-third of the German farmers are actually full-time farmers. And only one-fifth of the 2 million persons of working age living on farms were male full-time farmers, i.e. of the category that is used as the socioeconomic yardstick in agricultural economics.

For official agricultural policy, part-time farming was mostly seen as an undesirable phenomenon. Also it was believed to be a transitory stage from full-time farming to the abandoning of farms. Only in recent years the official attitude became more realistic and positive towards part-time farming. In the context of the CAP reform part-time farming may not only be stabilized but even become an important element of the reform. Part-time farmers are less vulnerable to environmental restrictions and to falling farm prices. They tend to be more qualified for the exploration of other income opportunities; states wanting to maintain the economic and cultural viability of the land under the present conditions of farm prices will have no choice but to actively create additional job opportunities for farmers or their families.

Among the possibilities are

1. agricultural counselling, e.g. for lower inputs of chemicals;
2. breeding and in situ conservation of rare crop strains and animal races (state subsidized);
3. winter services, e g. for roads, game feeding, etc.;
4. tourism, well established in mountainous areas and at sea shores but also a growing opportunity near the cities (weekend-tourism);
5. rural restaurants offering home production; some may find attractive specialization, like game and fish, health food or gourmet cuisine;
6. food processing;
7. nature protection (see Sect. 2.8)
8. crafts and trades;
9. industrial work, facilitated by new decentralization possibilities;
10. services including health care services e.g. for long-term patients; secretarial, administrative and related services can also increasingly be decentralized.

As more and more farmers seize such opportunities also the rather negative image of part-time farming is likely to disappear and even to be reversed. The main asset of rural areas, comparatively healthy living conditions, should be made an important element in promoting modern part-time farming.

7.2.13 Hobby Agriculture

Hobby agriculture is an extreme form of part-time agriculture, mainly in Northern European countries. It is clearly a phenomenon of affluent societies and is bound to spread as prosperity grows.

In the past hobby farming was both quantitatively marginal and undesired by the farmers lobby. Also this may change over time.

Hobby farmers are even less vulnerable to environmental restrictions; in fact they are likely to be eager to do even more than required. Their motive is not the maximization of production but pleasures of work and healthy food for the family and perhaps a few friends. In the context of the CAP reform certain hobby farmers may be willing to serve as 'guinea pigs' for necessary experimental exploration. The economic impact will, however, remain very limited.

7.2.14 Organic Farming

For a long time organic farming was just a craze of a few intellectuals and traditionalists. During the last ten years, however, it has become a serious activity notably in France and the northern EEC countries. Scenarios have been proposed even for converting all European agriculture into organic agriculture within 45 years (Bechmann 1987: 170). Also the agricultural scientific community is becoming increasingly interested in assessing the differences among different farming methods.

A certain controversy remains about the meaning of the word. To offer two extreme interpretations: If it is just another word for sustainable agriculture, organic farming is nothing less than a necessity. If it denotes, however, the specific meaning of Rudolf Steiner's 'biological-dynamic' farming, it is bound to remain restricted to small circles, or a 'craze'.

Assuming that under all definitions organic farming is ecologically preferable to conventional methods the question remains nevertheless if organic farming can play a significant role in the CAP-reform. For the German Greens and related groups the situation is clear: the state should guarantee even higher prices than today for all small farmers (not for the hated big ones) and to finance the transition into organic farming of all farmers. Assuming that such proposals are seen as outlandish in the present discussion, other forms of state support have to be discussed.

One important element was mentioned in Sect. 2.1: permitting and even encouraging Europe-wide labels for organic food that may fetch higher prices. Proposals do, in fact, exist. Further, certain limited subsidies to cover some transition costs into non-polluting farming are no less justified than the corresponding compensation payments mentioned in Sect. 2.6. Finally, the state funded research, so far almost exclusively concentrated on productivity innovations, should much more seriously and comprehensively address questions of feasibility, reliability and quality of organic farming. It should be noted that some of the other elements mentioned in this essay have an implicit supportive effect for organic farming. There are e.g. organic farmers who claim that the variety of birds and butterflies is by a factor 3 higher than on neighbouring conventional farms; if this is true, premiums for species variety would go preferentially to the organic farmers.

7.2.15 Direct Marketing

Farmers can become quite angry when seeing that the consumer normally pays twice, five times or even ten times the price the farmer gets for his products. Evidently the difference is caused by the costs of transport, processing, packing, preservation, marketing, hygiene controls and other factors.

Direct marketing avoids most of these costs and brings more cash to the farm. But there are many obstacles to direct marketing, not only the geographical distance from the consumer. To a large degree it is also an exaggerated hygiene control and EEC-wide standardization that has made it difficult for farmers to sell their products even to the local food store. Where ever the consumer is prepared to assume a certain risk, the state should not prescribe unnecessary standards. More flexibility in this domain may be a prerequisite for the development of self-sustaining local agricultural production that is infinitely less costly to the tax payer than the present system and is likely to be much healthier overall.

7.2.16 Food Security Policy and Active Price Policy

The CAP consisted up to now mostly of an 'active price policy'. This is precisely what the CAP reform will have to change. However, there may be good reasons for maintaining at least certain aspects of the policy of artificially high prices. In Switzerland, Norway and Finland a high price regime is maintained in order to keep farming activities going in these climatically less favoured countries. The chief reason for this is precaution for possible days of crises. In case of a collapse of international traffic streams or of other civilization support systems any country living almost exclusively from imported food would run into a major famine. If the possibility of catastrophic crisis is taken seriously, not only food imports will be affected. Fuel for heavy machinery, electricity for milking machines and the logistics of large scale slaughterhouses may equally fail.

Hence, a food security policy would require the maintaining of a minimum level of local low-input farming. It can be argued that this can be best achieved by a high price system, as the German Greens suggest, exclusively for small farmers, but other mechanisms can equally be explored. It should be noted in any case that low-input farming, part-time farming, hobby farming and functioning direct marketing systems are all elements of food security. Also stable and healthy ecosystems constitute an element of resilience for possible days of crisis.

It is clear that these 16 elements are still rather selective. Many purely economic instruments of the current CAP debate were left out, such as the quota systems and the co-responsibility levy but this essay has the ambition only to outline those factors which are of major ecological significance in the context of the CAP-reform.

7.2.17 Preliminary Assessment of the Package

For an assessment of the package three questions will have to be answered:
(1) what are the financial implications?
(2) how will the different actors react?
(3) how does it affect the environment?

The third question is easy to answer: nearly all elements are designed to have a positive effect on the environment. Questions (1) and (2) will be dealt with in a preliminary fashion in the following subchapters.

7.3 Financial Implications

None of the sixteen elements has been presented in a form allowing a financial assessment in quantitative terms (which would be needed country by country and for the EEC as a whole). The ambition of this preliminary assessment is only to give a sense of direction for each element and, thereby, for the whole package. As will be said in the final subchapter: Both for farmers (defending the present CAP) and for reformers (wanting to reduce the burden and to improve the conditions for the environment) there will be a trade-off between different elements. This should become more visible in the following very crude tabulation listing all sixteen elements and estimating their effects for farmers and taxpayers and for the environment.

To summarize the results from this preliminary assessment:

The package seems to contribute significantly to the reduction of agricultural surpluses.

1. For farmers the balance would not be negative (although *some* who are specifically benefiting from the current surplus production will certainly suffer losses, while others, rather small farmers living not too far from cities are likely to benefit). In view of the desperate prospects for many farmers if the present CAP is dragging on under steadily deteriorating conditions, a non-negative balance of the package is in fact an unexpectedly positive result.
2. For the environment the package would be a big improvement.
3. Taxpayers are likely to gain. A true balance can only be expected once all elements are available in quantitative form.
4. There are losers in the game: agrochemical industries, store houses, feed trade, farm machine industries and some industrialized farms. However, there are also interesting business opportunities lying in the ecological transformation.

7.3 Financial Implications

5. There are some positive side effects notably in public health and for water resources.

The contentious question of lowering EEC guarantee and intervention prices to world market levels was not explicitly raised in this essay. But if surpluses are dramatically reduced, the question loses much of its explosive character. Conversely, if farm incomes from other sources rise, the further lowering of guarantee prices loses much of its threat to the farming community. Hence it seems that the package would indeed contribute to a solution in the required direction.

It is obvious that further quantitative studies will be needed before any final assessment of the financial impact can be done.

7.3.1 Political Acceptance

All elements are well known and for all there is at least some practical experience available. All have their supporters and their adversaries. Environmentalists would opt for elements No. 1–8 and possibly Nos. 9 and 14. Farmers would not mind limiting the discussion to elements No. 16 and, more reluctantly, Nos. 8–15. And fiscal authorities are happy with all environmental demands leading to a reduction of surpluses.

Moreover, there is the notorious international dimension: For France anything reducing the stream of ECU's into her farms will be looked at with suspicion, while Britain will evaluate the package under the criterion if it is suitable to put the stopper into the bathtub of EEC finances (to use the British Prime Minister's expression after the Brussels summit). Mediterranean countries will watch for consistency with EEC regional policy, while Holland, Denmark and North Germany will try to rescue as much as possible for their intensive farms.

No.	Element	Surplus reduction	Effects on farm income	Ecological effects	Effects for taxpayer	Losses for others?	Other benefits
1	Food quality labels	+	+?	+	+	Chemical industry	Health
2	Biocides restriction	++	– –	++	++	Chem. ind. consumers[a]	Health water
3	Intens. livestock limit	+	–	+	+	Feed trade pharm. ind. consumers[a]	'Ethics'
4	Nitrates control	+	–	++	+	Nitrates chem. ind. consumers[a]	Water health
5	Applying env. regulation	+	–	++	+	0	0

(continued)

(continued)

No.	Element	Surplus reduction	Effects on farm income	Ecological effects	Effects for taxpayer	Losses for others?	Other benefits
6	Compensation payments	0	+	0	−	Water consumers?	0
7	Extexsification schemes	+	+[b]	++	−/0	0	0
8	Payment for conservation work	0	+	++	− −	0	Water resources
9	Set aside	+	+[b]	+	+	0	Forests
10	Early pension schemes	+	+	0/−	0/+	0	0
11	Biotechnology	+/−	+?	+/−	+/−	0	'Technical' progress
12	Part time agriculture	+?	+/++	+/0	+	0	Modernis. rural areas
13	Hobby farming	+	0	+	+	0	Quality of life
14	Organic farming	+	+	++	+	0	Health
15	Direct marketing	+?	+	0/+	0	Trade health risks?	Health?
16	Food security policy	−	++	−?	− −	0	National security
Package 1–16		++	0	++	+	Store houses chemical industry	Health natural resourses

[a] These elements may lead to rising food prices
[b] Compensation will lead to be lower than the income losses

As in all real political situations there will be scope for bargaining. Progress in such negotiations has three prerequisites: urgency of the situation, strong promoters and a sense of fairness. The urgency of the CAP reform is obvious. Promoters of a liveable compromise are abound; and this essay tries to hammer it into the environmental groups that they should forcefully join the lines of the promoters. Regarding fairness this essay tries to say that all sixteen elements form a package and that one should not insist on any particular sector out of the package without acknowledging the merit of the other ones.

References

Baldock, David; et al. 1986: *Agriculture and Environment: Management Agreements in Four Countries of the European Community* (Luxemburg: EU).
Baldock, David; Conder, David, (Eds.), 1988: *Removing Land from Agriculture—The Implications for Farming and the Environment* (London: IEEP) (This volume forms part I of the study commissioned by the German BMU of which the present essay forms part II).

References

Bechmann, Arnim, 1987: *Landbau-Wende, Gesunde Landwirtschaft—Gesunde Ernährung* (Frankfurt: Fischer).
European Parliament, 1986: Resolution "Agricultural Environment", 19 February 1986, Doc. A 2-207185 (The French text of the resolution is printed in Francois Roelants du Vivier: *Agriculture europ6enne et environnement*, 1987).
Hagedorn, Konrad; Klare, Klaus; Wilstake, Lugder, 1986: *Flächenstillegung mit Vorruhestandsregelung als Soziales Marktentlastungsprogramm: Ausweg oder Irrweg?* (Völkenrode bei Braunschweig: Bundesforschungsanstalt).
Hötzel, Hans-Joachim, 1986: *Umweltvorschriften für die Landwirtschaft* (Stuttgart-Hohenheim: Ulmer Verlag).
IEEP—Memorandum, 1987: "Biotechnology and the Environment in the European Community", written evidence to the British House of Lords, February 1987
Lavoux, Thierry, 1987: *Impacts sur l'environnement des biotechnologies—risques et opportunités* (Bonn: IEEP).
Priebe, Hermann, 1985: *Die subventionierte Unvernunft, Landwirtschaft und Naturhaushalt* (Berlin: Siedler Verlag).
Priebe, Hermann; von Weizsäcker, Ernst Ulrich, 1987: *Zur Neuorientierung der EG-Agrarpolitik unter umweltpolitischen Gesichtspunkten*, Bearbeiter: Otmar Seibert, Heino von Meyer, Untersuchung im Auftrag des Hessischen Ministers für Landwirtschaft, Forsten und Naturschutz (July 1987).
Rexen, F.; Munck, L. 1984: *Cereal Crops for Industrial Use in Europe*, Report prepared for the Commission of the European Communities (Brussels: EG, DG XII): 9617.
Roelants du Vivier, Francois, 1987: *Agriculture europ6enne et l'environnement—un avenir fertile* (Paris: Editions Sang de la Terre).
SRU (Der Rat von Sachverständigen für Umweltfragen), 1985: *Sondergutachten: Umweltprobleme der Landwirtschaft* (Stuttgart: Kohlhammer).
Sukopp, Herbert; Hampicke, Ulrich, 1985: "Ökologische und ökonomische Betrachtungen zu den Folgen des Ausfalls einzelner Pflanzenarten und—gesellschaften", in: *Schriftenreihe des Deutschen Rates für Landespflege*, Heft 46, August 1985: 595–608.
von Urff, Winfried; von Meyer, Heino, (Eds.), 1987: *Landwirtschaft, Umwelt und ländlicher Raum—Herausforderungen an Europa. Hermann Priebe zum 80. Geburtstag* (Baden-Baden: Nomos): 273–296,
von Weizsäcker, Ernst, 1988: "Anmerkungen zur Freisetzung gentechnisch veränderter Organismen in die Umwelt", in: *NATUR*, January 1988.

H. E. Secretary General of the United Nations, Ban-Ki Moon, shaking hands with Ernst Ulrich von Weizsäcker during a dinner on 30 January, 2014, of the German Association for the United Nations (DGAP) where the author gave a brief dinner talk. *Source* DGAP, with permission to use

Chapter 8
Earth Politics

8.1 Introduction: Why Earth Politics?

The Earth Summit has been and gone. With more than a hundred heads of state or heads of government present, it was easily the largest diplomatic event of the century. The summit was devoted to the environment and its links to development, or at least that is how the Northern media presented it. From a Southern perspective, the United Nations Conference on Environment and Development (UNCED), held at Rio de Janeiro from 3–14 June 1992, was devoted to development, global inequalities and their links with the environment. The connections between the environment and development are so close and intricate that any attempt to disentangle them can only lead to illogical conclusions and counter-productive solutions.[1]

To show this, let us start by looking at the Southern myopia, which is easy for Northern readers (who probably constitute the larger part of this book's readership) to criticize. The South has tended to define the UNCED agenda as a strategy for overcoming poverty and reversing global economic inequalities. Environmental questions would in this interpretation be subordinated to global economic issues. This view almost invariably leads to a repetition of Indira Gandhi's famous statement 20 years ago at the Stockholm UN Conference on the Human Environment, that 'poverty is the biggest pollution'. That may well be the case in areas where poverty drives people to collect and cut firewood in an unsustainable manner, and in a different sense it is also true to say that poverty under present conditions means high birth rates and resulting population pressure which then leads to further environmental degradation.

On the other hand, overcoming poverty in the conventional sense tends further to increase the stress on the environment. It means higher per capita consumption

---

[1] This text was first published in German in my book: *Erdpolitik. Ökologische Realpolitik an der Schwelle zum Jahrhundert der Umwelt* (Darmstadt: Wissenschaftliche Buchgesellschaft). The English edition, translated and edited by Ulrich Loening, University of Edinburgh, was published as: *Earth Politics* (London: Zed Books, 1994). Permission to republish this text was granted on 17 February 2014 by Renata Kasprzak, Rights Manager, Zed Books, London.

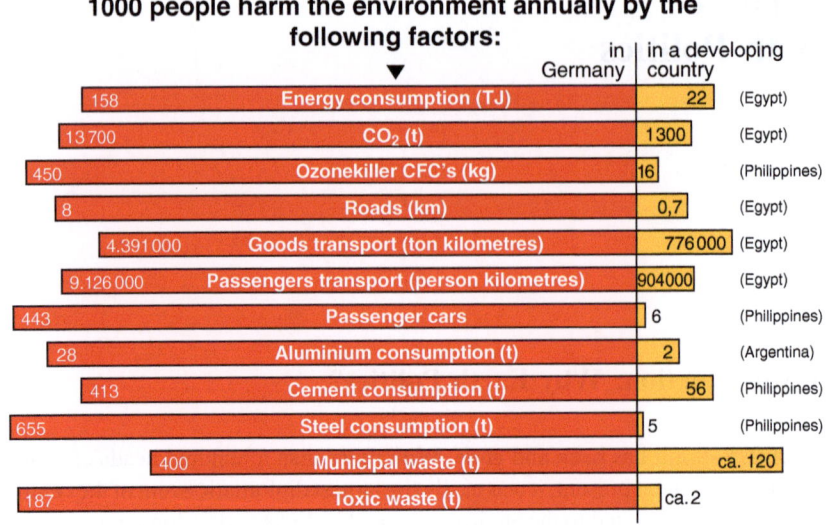

Fig. 8.1 People in the North consume roughly ten times more resources than people in the South. Figure by Hans Kretschmann, Wuppertal institute.

of water, energy, biomass, minerals and higher rates of land use, as well as waste, water and air pollution. Reversing global inequalities does not seem remotely possible in the real world without a massive increase in worldwide consumption rates. Hence giving priority to issues of development and world equity means that the environment is doomed to deteriorate at an accelerated pace.

The Northern myopia is no less fallacious. It essentially consists of saying that environmental protection is of the highest priority (together with birth control—in the developing countries, of course!). Environmental protection, on the other hand, is a high-tech, costly activity which even the North can afford only under conditions of affluence. The South is advised to strive for economic growth (through world market integration and activating the private sector) so as to be able to afford costly Northern pollution-control technologies. As a compromise during an interim period, some official development assistance may be made available for this purpose. This is how the North saw the negotiations at Rio de Janeiro for Agenda 21, that vastly comprehensive programme of action on all conceivable areas of environmental protection.

But what is meant by affluence? It is essentially the Northern way of life with a healthy diet, high levels of individual mobility, space for living, infrastructure, education and so on. Figure 8.1 shows that present lifestyles in Germany are associated with consumption rates per capita roughly ten times higher than those in some developing countries. If such German, let alone North American, levels of consumption were extended to some 5.5 billion people, the Earth would in no time be completely exhausted, swamped and overheated, in short destroyed.

8.1 Introduction: Why Earth Politics?

Caricaturing the Southern and Northern myopias in this way we have also outlined the dilemma faced by the Earth Summit, and indeed the World Commission on Environment and Development at its establishment in 1984. The World Commission, commonly referred to as the Brundtland Commission after its chairwoman Dr Gro Harlem Brundtland, Prime Minister of Norway, has given prominence to a formula, 'sustainable development', to overcome both the Southern and the Northern myopias (World Commission 1987). This formula alone, as everybody knows, does not and cannot change the world. [...]

What then can change the world? To escape the dilemma of Northern and Southern myopias and to make sustainable development a reality, we need a new set of values, a new culture, a new set of incentives to make those millions of actors both in the North and the South act differently. We need to enter and to create a new era of human history. The Earth Summit left us with this challenge. I call it the challenge of Earth Politics.

The Earth Summit has served as a good start for Earth Politics, but more fundamental changes are due. In my view, we shall have to overcome the obsessive way in which we subject everything on Earth to the economy. Economics is the common denominator of the Southern and the Northern myopia. Economics has become the veritable religion of our time.

8.1.1 *The Economic Century: An Episode*

Each century in modern times has had its own particular character. Our century is the Economic Century. Realists, or those who consider themselves realists, base their actions on economics or use economics as a justification for those actions. The world is divided according to economic criteria. Nowadays it is more important to know whether countries are (economically) 'developed' or 'less developed' than to know about their climate, their government or their religion. The European Union was, of course, founded as an economic community. The upheavals in Eastern Europe were, at least in large part, brought about by the hope of economic change. Worldwide, elections are usually decided by economic factors, and most activities of the United Nations agencies and organizations relate to the economic development of the developing countries.

Science and technology, which once belonged intellectually in the realm of the arts, have become crucially important areas of economic activity. Meanwhile, the arts themselves have also become an important sector of the economy. The Beatles were honoured by the Queen for their economic success rather than for what she or the Prime Minister considered to be their artistic merit. Even Christmas is nowadays first and foremost an economic event: it's the day after the last shopping day before Christmas.

In the second half of our century, economics holds us in thrall to such an extent that it never occurs to us that other centuries were shaped by entirely different realities or perceptions of reality. One only has to read the literature and records of past centuries to discover that, contrary to our own economic prejudices and to the Marxist interpretation of history, the economic worldview hardly ever played a dominant role in our culture before 1900.

In the 17th century—at least in central Europe—religious wars determined our culture. To be a realist in the 17th century meant to prove oneself worthy of one's denomination. The Counter-Reformation was an ecclesiastical, state and cultural response to the challenge of the Reformation, which itself had shaped the 17th century. According to prevailing public perceptions, these religious developments had virtually nothing to do with economics.

The 18th century became the century of the princely courts. As a realist you kept on good terms with your Prince. Soldiers served the King. Even the Enlightenment, which sowed the seeds for the end to royal absolutism, originated in courtly circles. Voltaire was a favoured guest of Frederick II of Prussia. To be sure, there were already some economists, the physiocrats in France and Adam Smith in Britain. The former, however, were but a passing phase and Adam Smith primarily considered himself a philosopher, although he also represented the advanced British culture of the time which was to discover economics much earlier than other cultures did.

In the 19th century the princely courts gave way to nation states created by bourgeois revolutions. A realist now defined him- or herself in France as French, in Germany as German, or in Britain as British. Garibaldi and Cavour were the Italian heroes of the time, the men who 'created the Italian nation', as old-fashioned historians would have put it. To be sure the 19th century may also be characterized by a phenomenon which is usually seen as an economic phenomenon, namely industrialization and the emergence of the working class. But workers and industrialists alike found their identity in their nation states. The workers' idea of an international community or solidarity remained a dream.

The European nation states soon discovered how to harness modem industrial and military technology to conquer the world (see Polanyi 1944, and Chap. 3). An ugly phase followed with imperialism, colonization, global economic crises, mass unemployment, totalitarianism and last but not least two devastating world wars. The age of the nation state, one should think, had come to a natural end. A new paradigm could move in: economics.

Contrary to our ethical traditions, egotism and the pursuit of short- to medium-term material and financial gain have been accorded a place of honour, through a simplistic interpretation of Adam Smith: the Invisible.

Hand of God would ensure that this preoccupation with the pursuit of individual profit would be to the common good. The broad consensus of the system of values to which we adhere in our present century—the century of economics—is based on this anti-ethic (see, e.g. Lux 1990).

It is easy to understand how the economic paradigm took hold so forcefully after the end of the Second World War. Economic thinking, so gloriously represented by the USA, the liberator of the devastated Old Continent, meant peace, freedom and the prospect of material welfare. International politics was to be redefined in terms of reconstruction and development. The very term 'underdevelopment' in the economic sense was publicly pronounced by President Harry Truman only in 1949 (see Oxford English Dictionary, vol. XVIII: 960).

World trade, development assistance, infrastructure and technological development: these were just a few of the catchwords of the new economic consensus of the Western world. The United Nations was inaugurated with more or less equal emphasis on security and economic development. At the Bretton Woods Conference in 1944 the new international economic system was defined, including the International Bank for Reconstruction and Development (the World Bank), the International Monetary Fund (IMF) and later the General Agreement on Tariffs and Trade (GATT). The goals of all these new institutions, as of the leading nation of the time, the USA, were peace and prosperity for all. And economics was to be the dominant moving force. Who could deny the spell of such a promise?

From the vantage point offered by so attractive a set of values, it is not difficult to see how the concepts which had shaped the cultures of earlier centuries came to be seen as unenlightened and foolish. But that may not do justice to the people who lived before us. And who knows whether people in future centuries will find our present day economic values and preoccupations even more foolish and unenlightened than the values of religious hypocrisy, princely courts or the nation state?

Indeed, such a judgement of history on our own time is precisely what I fear. Many critics have pointed out that under the veil of seemingly non-political economic values, oppression, conquest and abuse of power continue. However, the real danger, as I see it, is that the supremacy of economics in its present form will cause irreparable damage to the Earth and to the people who live on it and from it, irrespective of the injustice and deprivation which are caused, or at least not prevented, by the dominion of economics.

In the parlance of present-day economics and politics, the realist thinks in the short-term and does for nature, the environment and posterity no more than the legally prescribed minimum. To do more means to commit oneself to expenditure without profit. In many cases it is even regarded as absurd idealism to follow the letter of the law. Indeed, on occasion the authorities in many countries shut their eyes because they are more concerned with short-term prosperity, jobs or tax revenues than with the environment.

8.1.2 The Rape of Nature

It is my thesis that the wonderful days of naive economic consensus are numbered. We are reaching the limits of destructive growth. It is just not possible for the amount of energy, land, water, air and other natural resources consumed by 10 % of the world's population—directly or indirectly—to be matched by the remaining 90 % without total ecological collapse. And yet this 'standard' is the declared aim or the dream of all development aspirations.

No invisible hand can ward off such an ecological collapse. Merely to sustain the present level of consumption of the top 10 %, natural resources are being exploited at a fearful rate. At present approximately 1,000 tons of soil are being washed, blown away or otherwise eroded per second; the Earth is losing some 3,000 m^2 of its forests per second, an area almost the size of Britain, each year; each day we wipe out ten, perhaps fifty, species of animal or plant; every second we pump around 1,000 tons of greenhouse gases into the atmosphere (Lester Brown et al. 1988, Table 1.1).

Ecological disasters on a local scale make things worse. In Mexico City and Wuhan (China), the air is so badly polluted that hardly a child grows up without suffering from chronic lung disease. The Ivory Coast in West Africa has lost three-quarters of its forests in 20 years. The Vistula River in Poland, once known for the quality of its fish, is now virtually dead and its water is considered unfit even for industrial use. The once forested Riesengebirge between Germany and Bohemia are now largely bare, except where it has not yet been possible to keep up with the felling of the dead or dying trees. The Baltic and the Black Seas are both under severe threat of devastating eutrophication (Alcamo 1992).

Worst of all, there are the global problems. The ozone layer, which largely screens out cancer-inducing ultraviolet-B-radiation, is being destroyed by man-made chemicals, chiefly chlorofluoro carbons (CFCs). Our global climate has been by and large stable during the present interglacial period, but now there are signs that, by geological standards, the climate is undergoing extraordinarily rapid change as a result of human activity, in a way which could turn whole regions of the earth, including parts of Europe, into steppe or desert and raise the ocean levels to an unpredictable extent (for a balanced reading, see Houghton et al. 1990). Uncounted additional species may become extinct if climatic change progresses at the speed being forecast and leads to an unprecedented shift of biotic zones (Peters/Lovejoy 1992).

Tragic situations sometimes begin in an invisible way only to become apparent when it is too late to attempt to solve them. William Stigliani (1988) gives the striking example of the build-up of acidity in the Big Moose Lake in New York State. For 80 years increasing quantities of sulphur in the form of sulphurous or sulphuric acids rained down on the region, until finally the buffering capacity of the soil in the catchment area and of the lake itself were exhausted. The last drops of acid rain eventually pushed the lake past the point of no return, since when it has been virtually dead.

A similar story can be told of Waldsterben, the dieback of forests which became apparent only a decade or two after the process had actually begun. And only now can we see the effect of nitrates which have been seeping into the groundwater for the last 10 or 20 years. There are other examples where, over large areas, the considerable ability of the soil to act as buffer has been similarly exhausted: we may already be sitting on hundreds of ecological time bombs threatening the livelihood of our children.

8.1.3 Earth Politics for the Century of the Environment

If the days of the Economic Century are numbered, what of the future? Whether we like it or not, we are now entering a Century of the Environment. In this new century, the hallmark of the realist will have to be regard for the environment. Short-term economic goals will naturally remain, but if they are not subordinated to the ecological imperative they will in time lose all credibility.

At first sight, a Century of the Environment seems an optimistic vision, but that is not what I mean to convey by the phrase. What it signifies is the cruel reality of ecological devastation which will confront us in our everyday lives and which will inevitably shape our civilization if present trends of destruction continue for just another decade or two. And, given the massive momentum behind present trends and the well-known slowness of human beings to change, there is no doubt that this pillage will continue for much more than two decades. This means that, from its very beginning, the 21st century will bear the mark of a massively endangered natural environment. This fact will become increasingly dominant in all fields of politics, from foreign affairs and development policy to research, technology and education. In the Century of the Environment it is the ecological imperative which will become the dominant determinant for law and administration, for city planning and agriculture, for the arts and for religion, for technology and indeed for the economy.

The transition from our present Economic Century to the Century of the Environment will not simply follow from somewhat more ambitiously formulated standards for water, air and soil pollutants, nor from symbolically upgrading the post of environment minister in governments throughout the world. Clearly, the transformation has to go much deeper than that. The sooner we make a start, the better our prospects of salvaging for future generations the amenities and the fabulous degree of freedom which, at least in some parts of the world, made the Economic Century so attractive.

The political task associated with this impending transformation is what I call Earth Politics. Earth Politics will have to be very pragmatic. It must take realistic account of the present and its power structure and it must not demand the impossible of either the people or the decision-makers. In order to evaluate the present accurately, Earth Politics must also show an understanding of the past. For this reason, almost every chapter of this book begins with a few words on the historical background. Not least, since its origins lie in the Economic Century, Earth Politics must take into account the contemporary political bias towards economics. It must develop and offer, so far as possible, economically viable strategies for the impending transformation.

Earth Politics must be international. We must get out of the habit of thinking in terms of the nation, a concept which had its heyday in the 19th century, but at the same time we must respect the human need for a sense of home and place, for linguistic and cultural identity, as well as for community. Simultaneously, much political decision-making should be decentralized: 'Think globally—act locally'.

For all its pragmatism, Earth Politics also needs a vision. That vision must be consistent. In particular it will also have to address, and as far as possible resolve, the fundamental contradiction between the level of consumption of today's rich and what is ecologically feasible for 5 billion, and 1 day 12 billion people.

The first part of the book seeks to set the ecological and historical framework for Earth Politics. It is worth paying attention to the beginnings of Earth Politics from within traditional environmental politics because it is here—most encouragingly—we discover that it is possible to bring about meaningful change in the economy in a relatively short space of time and without needing a revolution.

The second part then examines five topics or areas of crisis in which classical environmental politics has not been successful; first steps towards solutions in these areas are outlined.

The third section moves on to put these fragmented elements in a coherent context, which is intended to serve as a politically realistic plan of action based on the generally accepted polluter-pays principle of environmental policy. However it is shown that, in spite of assertions to the contrary, classical environmental policy has at best been only half-hearted in translating this principle into reality. This is the basis from which the concept of ecological tax reform is derived and a new direction for foreign policy after the Earth Summit may evolve.

Finally, the fourth section tries to give clearer expression to the vision of Earth Politics, though it can hardly yet lay claim to being regarded as pragmatic or constituting realpolitik. A longer-term vision is nevertheless required to impart the sense of direction which is so deplorably lacking in present-day realpolitik.

It is to be hoped that this whole construct—the framework, areas of crisis, a plan of action and the vision—will appeal to the consciousness of the reader—the consciousness that 'things can't continue like this', but also that we still have a good chance to set the necessary changes in motion. Such consciousness is crucial for the earth-political change which we now need. [...]

When prices do not tell the ecological truth manufacturers and consumers are able to divert a goodly proportion of the real cost burden elsewhere. This diversion is known in economics as 'externalising' and results in external costs.

The 'ecological truth', of course, will never be precisely quantifiable in any scientific manner and we must remain aware of this basic fact in the discussion which follows.

Lutz Wicke (1986), a renowned author in environmental economics (and also deputy environment minister of the city state of Berlin), estimated the external costs arising from environmental degradation in West Germany approximately DM 100 billion based on 1985 prices, which represented over 5 % of that year's GDP.

Wicke studied only the domestic economic effects on air, water, soil contamination and noise. If we add external costs arising from accidents and environmentally caused illness, soil erosion, climate change, biodiversity losses and for damage exported or imposed on future generations, it does not seem difficult to justify a total of external costs caused by the West German economy in the order of DM 200 billion or some 10 % of GDP. In fact, Prognos (1992), the Basel-based research institute, in a study for the German ministry of economic affairs, assessed external costs of energy use alone at some DM 500 billion (this time including East Germany). On a global scale, Barbir et al. (1990) believe that external effects from burning fossil fuels alone may be in the vicinity of 14 % of the global GDP, an unimaginably high amount. [...]

As stated earlier, prices are a long way from telling the ecological truth. We are now in a position to give this assertion some quantitative meaning, bearing in mind that value judgments are involved in the figures. Today's costs, and therefore also today's prices, represent between only a quarter and a tenth of the 'ecological truth'. Put in a different way, if the costs of resource use and environmental pollution were increased by a total amount of approximately 5–10 % of GDP, then market forces could be assumed to work reasonably well to protect the environment. The market, the old culprit according to the early conservationists, would become their ally. [...]

8.2 Ecological Taxes and Charges

The traditional purpose of taxes is to raise fiscal revenue, but environmental taxes have a directional purpose in addition to that. Where environmental taxes are intended merely as a replacement for other taxes, rather than increasing state income, that directional purpose becomes dominant and we can talk of ecological tax reform. The revenue from environmental taxation is not controlled by the environment minister but by the finance minister, who is likely to veto any hypothecation. Not to veto it would be to limit the fiscal autonomy of parliament. It follows that environmentalists and environment ministers, seeing the need for ever greater funding for their multifarious activities, typically are not interested in environmental taxes, but rather in hypothecated charges or trust fund taxes the revenues of which are 'theirs'.

That may well be the reason why it took so long for an ecological tax shift to be introduced, for who, other than environmentalists, would be in favour? One could hardly count on Treasury politicians, for they are unlikely to espouse a tax whose express intent is to narrow the basis of tax revenue.

Not until environmental politicians are convinced that ecological tax reform would produce decisive, additional and directional effects will environmental taxes be adopted in addition to special charges or trust fund taxes. Assuming that they are targeted correctly, the steering effect of both charges and taxes chiefly depends on the level at which they are set.

The highest attainable level, or maximum extent of the directional effect, is where the decisive difference between hypothecated charges and ecological tax reform comes in. We established in the last chapter that the level at which charges can be set almost unavoidably remains very limited. Even during a period of high political acceptance of charges, their total amount barely reached a tenth of 1 % of GDP. There are other, economic-political grounds for this constraint on charges. For the economy, every special charge or additional trust fund is an additional

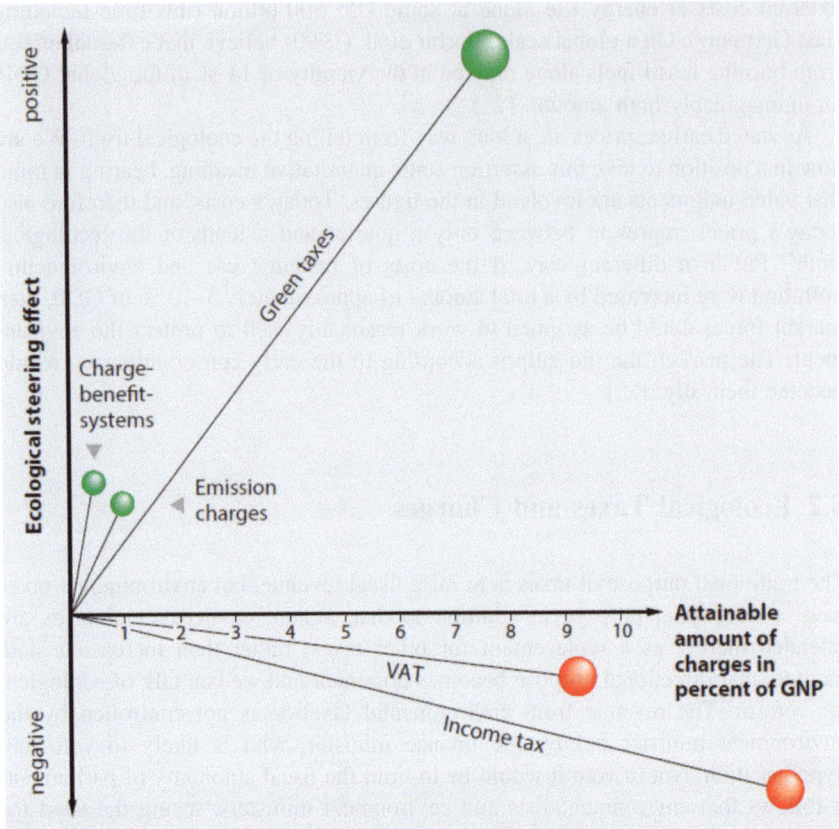

Fig. 8.2 The ecological steering effect of green charges (or hypothecated taxes) and ecological tax reform. The directional or steering effect follows the upward axis, and the precision of that effect (the specific effect per $ or £ of charge) is shown by the steepness of the gradient between zero and the respective instrument. Hypothecation is assumed to produce a high directional precision, ecological tax reform a comparatively low one. However, the total steering effect of environmental taxes within the tax reform will outstrip that of special charges provided taxes are raised in a revenue neutral way and thus achieve a substantially higher level than special charges. The positive steering effect of environmental taxes is reinforced if taxes with a negative ecological steering effect are reduced

burden. The business world is extremely suspicious of instruments which increase the fiscal burden and create new bureaucracies (as is intended in the case of special charges or hypothecated taxes). It is very unlikely that any government will ever get away with special charges at anything approaching the 5–10 % of GDP which we considered to be the external cost of resource use and environmental degradation. Let us assume, very optimistically, that the total revenue from special charges can attain five times its present maximum size, i.e. 0.5 % of GDP. This is the assumption underlying Fig. 8.2.

Environmental taxes within an ecological tax reform, by contrast, can grow to at least fifty times the order of magnitude of present special charges. Even a factor of 100 is legally possible and economically justifiable. A fifty- to one-hundredfold-increase in the weight of the signal would surely bring about a disproportionately faster and deeper reorientation of the whole economy. Both VAT and income taxes work as an added, macro-economically undesired incentive to replace human labour by machines and energy, thus increasing pressure on the environment. The positive effect of environmental taxes is therefore strengthened by the concurrent reduction of these taxes.

There are two principal reasons for allowing green taxes to reach a much higher level than special charges.

Firstly, the rate and level of taxes, unlike special charges, may be politically determined and are not tied to calculations of the costs associated with environmental damage or its restitution. The finance minister is under no obligation whatsoever to prove a quantitative relationship between the tax and damage to the environment. (Imagine a finance minister working under the obligation to first prove damages done by professional labour before being entitled to collect income taxes....)

Secondly, because environmental taxes within an ecological tax reform flow directly to the Treasury (and not to a trust fund), parliament can (and should) decide to reduce other taxes proportionately. There is then no increase in the average fiscal burden. Revenue neutrality is of crucial political importance and distinguishes ecological tax reform from the introduction of additional green taxes.

References

Alcamo, Joseph (Ed.), 1992: *Fair Wind, Foul Wind: Coping With Crisis in Eastern Europe's Environment*. (Laxenburg, Austria: IIASA).
Barbir, F; Veziroglu ,T; Please, H.J, 1990: "Environmental Damage Due to Fossil Fuel Use". in: *International Journal of. Hydrogen Energy*, 15: 739–749.
Brown, Lester, et al., 1988: *State of the World* 1988 (New York: W.W. Norton).
Houghton, R.A.; Skole, D.L.; 1990, "Carbon", in: Turner, Clark; Kates, Richards; Matthews, Meyer (Eds.): *The Earth as Transformed by Human Action* (Cambridge UK: Cambridge University Press): 393–408.
Lux, Kenneth, 1990: *Adam Smith's Mistake: How a Moral Philosopher Invented Economics and Ended Morality* (Boston: Shambala).
Peters, Rob, Thomas Lovejoy, 1992 (Eds.), *Global Warming and Biodiversity* (New Haven: Yale University Press).
Polanyi, Karl, 1944: *The Great Transformation: The Political and Economic Origins of Our Time* (Boston: Beacon Press).
Prognos, 1992: *Externkosten der Energieerzeugung. Gutachten für den Bundesminister für Wirtschaft* (Bonn: BMW).
Stigliani, William, 1988: *Changes in Valued Capacities of Soils and Sediments as Indicators of Non-Linear and Time-Delayed Environmental Effects* (Laxenburg, Austria: IIASA).
Wicke, Lutz, 1986: *Die ökologischen Milliarden. Das kostet die zerstörte Umwelt und so können wir sie retten* (Munich: Kösel).
World Commission on Environment and Development, 1987: *Our Common Future. The World Commission on Environment and Development* (Oxford–New York: Oxford University Press).

Chapter 9
Ecological Tax Reform

Ernst Ulrich von Weizsäcker and Jochen Jesinghaus

9.1 The Size of the Challenge

The *Intergovernmental Panel on Climate Change* (IPCC) calls for a major reduction, perhaps by 80 %, of greenhouse gas emissions a necessity if we are to finally stabilize the global climate before a catastrophic warming may have taken place.[1] CO_2, resulting mostly from the combustion of fossil fuels is the most important greenhouse gas (IPCC Intergovernmental Panel on Climate Change 1990a, b, c). An enormous gap is opening between the reduction needs and the business as usual increase of CO_2 emissions.

To try and fill the gap by nuclear energy appears a dangerous path and illusory too. It's not only nuclear safety that troubles people. Even the safest of today's reactors remain vulnerable to war action and sabotage and hence lend themselves as targets for blackmailing. And the safe handling of very large quantities of bomb-suited fissionable material that would inevitably appear on the market place is a nightmarish undertaking too. Thus the contribution of nuclear energy will remain modest.

Fusion energy is unlikely to be available in significant quantities during the next few decades, and the hazards may well turn out to be of the same order of magnitude as with nuclear fission power: There are the problems of enormous neutron flows bombarding and radio-activating any conceivable shielding material and the production of huge quantities of tritium, one of the most insidious radioactive substances.

Solar energy and other renewables have a better perspective from the technical and environmental point of view. And in certain places they have become competitive in recent years.[2] However, their general commercial viability remains limited under the prevailing world market conditions. Also they will probably be seen as environmentally damaging as soon as they are produced by the Gigawatts.

---

[1] This text was first published as in a book by Ernst Ulrich von Weizsäcker and Jochen Jesinghaus (1992). Permission to republish this text was granted on 17 February 2014 by Renata Kasprzak, Rights Manager, Zed Books, London.
[2] See Deudney/Christopher (1983), O'Keefe/Pearshall (1988).

All this means that for many decades to come fossil energy will most likely continue to represent the largest part of the energy pie. What is valid for world energy consumption is, *mutatis mutandis*, equally so for water use, materials consumption (on account of the impacts of mining, processing, transportation, manufacture and waste products), and ecologically detrimental land use. The gap between development 'requirements' under even modest assumptions and ecologically sustainable levels of consumption is growing ominously wide.

As long as the countries of the North persist in consuming around ten times more natural resources per capita than the developing countries, the gap will continue to grow wider and wider with the situation becoming increasing hopeless. A drastic reduction in per capita resource consumption in the North is ecologically imperative. And for the developing countries, it is a path of resource conservation—one that avoids the detour of the throw-away society—that will lead to the modern Promised Land. [...]

The United Nations in 1984 established the World Commission on Environment and Development under the leadership of Norwegian Prime Minister Gro Harlem Brundtland. The Commission's final report, *Our Common Future*, became one of the most important documents of the eighties. The Brundtland Report was very well received by the United Nations General Assembly which resolved to convene the *United Nations Conference for Environment and Development* (UNCED).

UNCED, often called the Earth Summit will be held in Rio de Janeiro in June 1992. It could well become the worlds largest international conference ever and might as well become one of the most important diplomatic events of human history. The challenge before the Conference is essentially the challenge outlined above: The [...] most important contribution by the North would be to change its own model. Sustainability, then, is primarily a task addressed to the North von Weizsäcker (1991). Our study therefore deals with the possibilities for the North of attaining a new and credible role model. The possibilities rest, so we argue, with a revolutionary increase of resource productivity.

To trigger off that new technological (and cultural) revolution we propose an ecological tax reform of which we are convinced that it can be tailored such that it is socially acceptable and would actually improve the opportunities for the business world. We think that the green taxes should primarily be levied on environmentally important input factors, notably energy. The taxes should lead to a steady price increase of 5 % annually over some 30–40 years for fossil and nuclear energy as well as for other problematic resources. Revenue neutrality could be secured by reducing other taxes such as VAT, income taxes or corporate taxes.

The imperatives stated above allow for one of two different approaches: Either a drastic reduction in per capita consumption, entailing sacrifices that would extend to voluntary poverty on the part of the North, or a drastic heightening of resource productivity. The first approach is obviously far less desirable than the second, and politically hopeless to boot, particularly when the second approach has yet to have even been tried. [...]

9.2 Ecological Tax Reform

An ecological tax reform is quite different from special, and earmarked ecological charges and from tradable permits. The idea is to put taxes on fossil and nuclear energy, water consumption, raw materials (especially those which are likely to end up as toxic pollutants or hazardous waste), or possibly emissions and waste, and to reduce other taxes instead. A revenue neutral tax reform would observe the stipulation that the overall fiscal burden must not increase. Revenue neutrality has a politically important terminological consequence: with revenue neutrality, one should not talk about ecological taxes, but rather about ecological tax reform. This is likely to meet with greater political acceptance than the term 'green taxes', which could be understood as representing an additional burden.

Unlike earmarked charges, the ecological tax reform requires no scientific proof of the causal chain lying between the taxed commodity or emission and the environmental damage. The Exchequer is never liable to prove that any of the factors which are taxed in our society causes damage. Of income taxes, VAT or corporate taxes nobody would even suspect that they are meant as a penalty against anything undesired. Rather human labour, the creation of added value and business activity are seen as something highly desirable for our economy. Thus, income and corporate taxes, just like the VAT in force throughout the EC, are seen by economists[3] as having a negative effect on the economy, albeit one which is generally accepted on account of the undisputed need of state expenditure. If there is no ecological burden of proof for environmental or 'green' taxes, their acceptance in the political arena nevertheless depends

- on their ecological plausibility, i.e. their expected capacity to steer in the intended direction,
- on social equity and
- on their effects, positive or negative, on the economy.

If ecological taxes are employed by the state in such a way that taxes with negative economic effects are reduced, and so that the overall fiscal burden is not increased, it could be expected that the impact on the economy might even be positive. The macroeconomic advantages are likely to be twofold: (1) less environmental damages, repair and health costs, (2) likelihood of increased employment because gross labour costs diminish as labour related taxes and charges are reduced.

Thus, even very high ecological taxes—provided that revenue neutrality is observed and that they were introduced slowly enough—could theoretically become acceptable by both industry and society. During the course of several decades, it ought to be possible to attain a level of 5–10 % of GNP. This is indicated in Fig. 8.2, where the estimated levels attainable by ecological taxes are contrasted with those for earmarked charges and charge-benefit systems.

---

[3] Cairncross, Frances, *Costing the Earth: What Governments Must Do; What Consumers Need to Know; How Business Can Profit.* London: The Economist Publications p. 96–99.

For environmentalists, the claim that ecological taxes could bring in as much as 5–10 % of GNP without damaging the economy as a whole would by itself be sufficient grounds for introducing an ecological tax reform. However, before the idea of embarking on such a programme of major social reorganization can be seriously considered, credible answers to a number of questions affecting both the economic and political spheres must be provided:

- What level or what order of magnitude of ecological taxation can be justified in terms of the polluter-pays-principle, even accepting that evidence of the precise origin of pollution or other ecological damage is not required?
- Does raising cost of resource consumption have the desired steering effect? And at what level is the steering effect relatively best?
- What is the optimum pace for introducing reform?
- To fulfil the requirement for revenue neutrality, which other taxes could best be lowered?
- How can undesired distribution effects be avoided or compensated for?

This study seeks to provide initial answers to these questions.

9.3 Price Elasticity

The next question asked is that of steering effectiveness. Even if an ecological tax reform is morally and theoretically justified, the question remains as to whether it will have the desired effect.

In casual political debates, the question is normally posed and answered in 'static' terms: How will the car user change his behaviour to reduce petrol consumption in reaction to a certain price increase or, in more scientific terms, what is the short-term price elasticity for petrol consumption or for any other product?

In reality that static or immediate elasticity is systematically much lower than the dynamic, long-term elasticity. And the elasticity to a short time price change is systematically lower than with a running, calculable price increase.

To clarify and illustrate the reasons for the discrepancy between short and long term elasticity we may distinguish five consecutive phases of adjustment to higher prices. The energy price movements in the seventies seem to provide a good empirical background. They were characterized by a shock-like price signal large enough to be noticed by everyone and to elicit immediate reactions.

9.3.1 First Stage of Adjustment

The consumer tries to get by on less energy. Hot water is no longer allowed to flow needlessly down the drain; windows are closed in winter and roller-shutters let down at night. Driving at top speed on highways and senseless accelerating in

9 Ecological Tax Reform

daily traffic will be reduced or avoided. Also, a number of minor sacrifices occur, e.g. dispensing with certain weekend trips by car, lowering the thermostat and doing without oil-heating for swimming pools.

This stage of adjustment affects energy consumption with practically no delay; however, the effects are limited (approx. 10–20 % during the second oil crisis of 1979–1980).

9.3.2 Second Stage of Adjustment

The criteria of energy efficiency become more important when purchasing energy-dependent goods (cars, household appliances) and functions (heating). Fuel consumption in cars plays a role again this process is also reflected in advertising: during the second oil crisis in 1979–1980, most advertisements emphasized fuel efficiency and explicitly provided fuel consumption data for speeds and conditions (90 km/h, 120 km/h, urban traffic). With the decline of prices, this sort of data largely disappeared from advertising and consumer test reports. This stage of adjustment takes effect gradually; the full impact is attained when all goods are replaced by more efficient ones.

9.3.3 Third Stage of Adjustment

Suppliers of energy consuming goods respond to the change in demand structure by developing more efficient types. In the housing construction sector, this means the development of more efficient heating systems and the reduction in heating requirements through more efficient insulation, control and air circulation systems. With cars, we could expect the further development of more fuel efficient diesel and Otto engines, as well as a reduction in wind resistance, tire friction and vehicle weight.

After running for a period of several years, this stage of adjustment produces distinctly tangible results. For instance, the low air resistance values of today's cars are the result of development efforts made in reaction to the high fuel costs during the oil crisis of 1979–1980.

9.3.4 Fourth Stage of Adjustment

The state and producers of energy consuming goods invest in the research and development of energy efficient systems and in technologies which are able to make do without fossil (and nuclear) energy. Examples of this would be innovative insulation and energy saving recycling systems, heat pumps, unconventional engines for cars (Stirling engines, gas turbines, and fuel cells), quick and efficient public transport systems with the requisite expansion of infrastructure, and finally the development of renewable energy sources.

This stage of adjustment attains its full impact only after an extended period (10–40 years), owing to the slowness of infrastructural changes and new technological developments, although certain existing results of research could go into serial production much earlier.

9.3.5 Fifth Stage of Adjustment

Demand for energy decreases due to a change in the housing structure, the infrastructure and the general way of life (culture, in the broadest sense of the word). For instance, the average distance between home and the work place could be reduced, leisure facilities could be relocated to locations closer to residential areas, and greenfield site shopping centres could be abandoned in favour of decentralized neighbourhood stores. Long-distance freight transport and the greater part of passenger-traffic could be shifted on to the by now more efficient rail system.

The full impact of this stage of adjustment will only emerge during the course of several decades.

The boundaries between each stage of adjustment are not clearly defined. It is important to acknowledge that the 'higher' stages require time. This is the reason why price-demand elasticity using short-term data only provide a false impression.[4]

A proper determination of the price-demand elasticity of energy could theoretically result if all of the above listed stages of adjustment were included, i.e. if the high price periods of the late 1970s had extended over several decades; however, during this period, other relevant variables would also have changed (e.g. per capita income, population density, and scientific and technical progress), which in turn would render causal analysis more difficult.

However, a criterion useful in determining long term price elasticity should be obtainable if price differences had existed over longer periods for certain identical or comparable goods in different countries with comparable economies. In this case, specific consumption differences for such goods could be correlated to the different prices. In doing so, the influence of other variables should, of course be considered but should be able to be separated from the price elasticity calculation.

In the next chapter, an example of a model for this will be discussed.

---

[4] See e.g. Sterner (1990). Sterner sets a short-term average price elasticity in OECD countries of 0.24 and a long-term price elasticity of 0.79!

9.4 Price Elasticity for Fuels: A New Measurement Concept

In the search for appropriate goods, and for goods which are relevant to the ecological tax debate, one quickly encounters automotive fuels. These have been very heavily taxed in Japan and Italy heavily but differently taxed in other West European countries, and hardly taxed at all in North America. They thus lend themselves to an initial assessment of long-term price elasticity better than virtually any other group of goods in an area of utmost relevance to environmental policy.

On the one hand, the purpose of such an assessment is to gain an indication as to how high ecological taxes have to be in order to guarantee that the ecological goals are successfully met; on the other hand, the assessment can provide an idea of the distribution effects caused by ecological taxes, and thus on the need for compensation measures.

Based on 1988 energy prices, we have conducted an initial analysis of the connection between the level of fuel prices and per capita fuel consumption in the most important OECD countries. It indicates—with just these two variables being employed—the existence of a high negative correlation, as is illustrated in Fig. 9.1.

The residual variance in relation to the regression line in the illustration is remarkably low. The correlation coefficient amounts to -0.93. This very high correlation may in part be due to coincidence, in part also to third factors correlating positively with fuel consumption and negatively with fuel prices. One such suspected factor is low population density which makes average distances longer and may make fuel taxation politically much more sensitive than in densely populated countries. We shall discuss this factor below. In a purely statistical sense, however, the negative correlation between price and per capita consumption is very well established. The relationship shown in Fig. 9.1 should make ourselves cautious about the widespread belief that the problem of fuel consumption can be dealt with effectively by the adoption of fuel efficiency standards. As is well known, the United States introduced car fleet efficiency standards in the mid-1970s. The graph shows that even after more than 10 years, these standards failed to pull the USA below the regression line. Some further reflection may offer an explanation for this failure: fuel efficiency standards have no restrictive influence on the number of kilometres driven; quite the contrary: at constant fuel prices they make more kilometres affordable to the car owners The best that can be said for the US fuel efficiency standards is that without them, the United States would probably be even worse off.in the international comparison.

9.4.1 Long Term Price Elasticity of Fuel Consumption is High

By its very nature, the straight line in Fig. 9.1 cannot realistically be extrapolated for higher fuel prices. This is because fuel prices that are only about a third higher

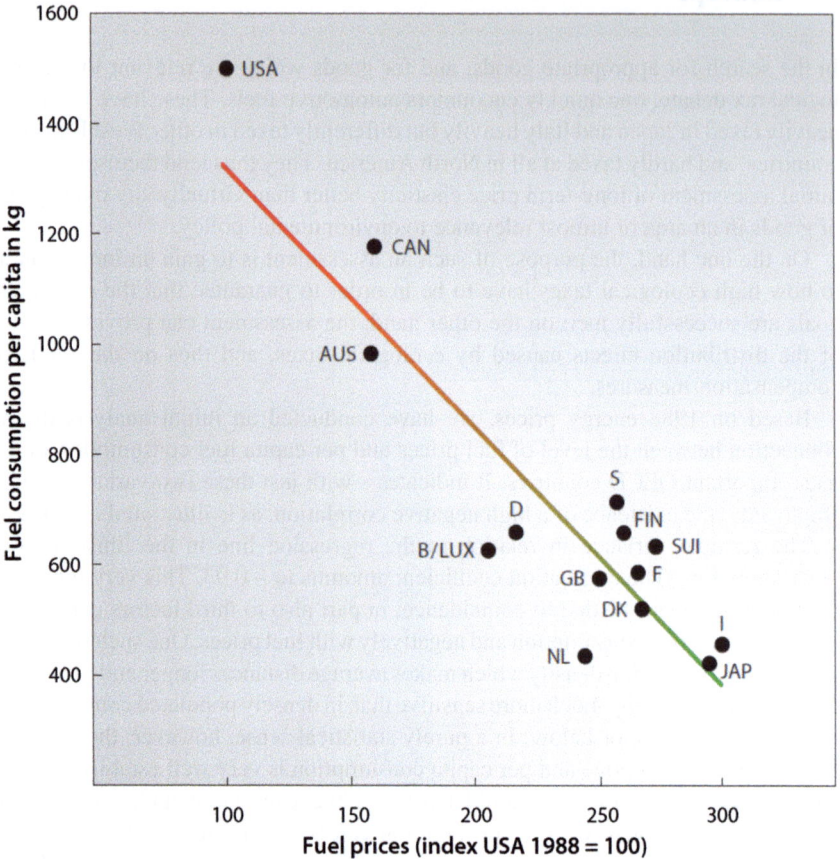

Fig. 9.1 Fuel prices and per capita consumption in the most important OECD countries, with letter-symbols corresponding to the standard abbreviations for the countries involved

than the Italian-Japanese level would result in negative fuel consumption rates. An asymptotic approach towards the zero line would instead seem plausible. A simple mathematical transformation will make it more convenient to extrapolate the findings beyond the Japanese price level. Instead of per capita fuel consumption we may plot its reciprocal value. The value obtained from the relationship 1/fuel consumption can be called 'macroeconomic fuel efficiency', (or just 'fuel efficiency'—where there is no confusion with the technical fuel efficiency applying to individual cars). Macroeconomic fuel efficiency is a measure of how efficient the

9 Ecological Tax Reform

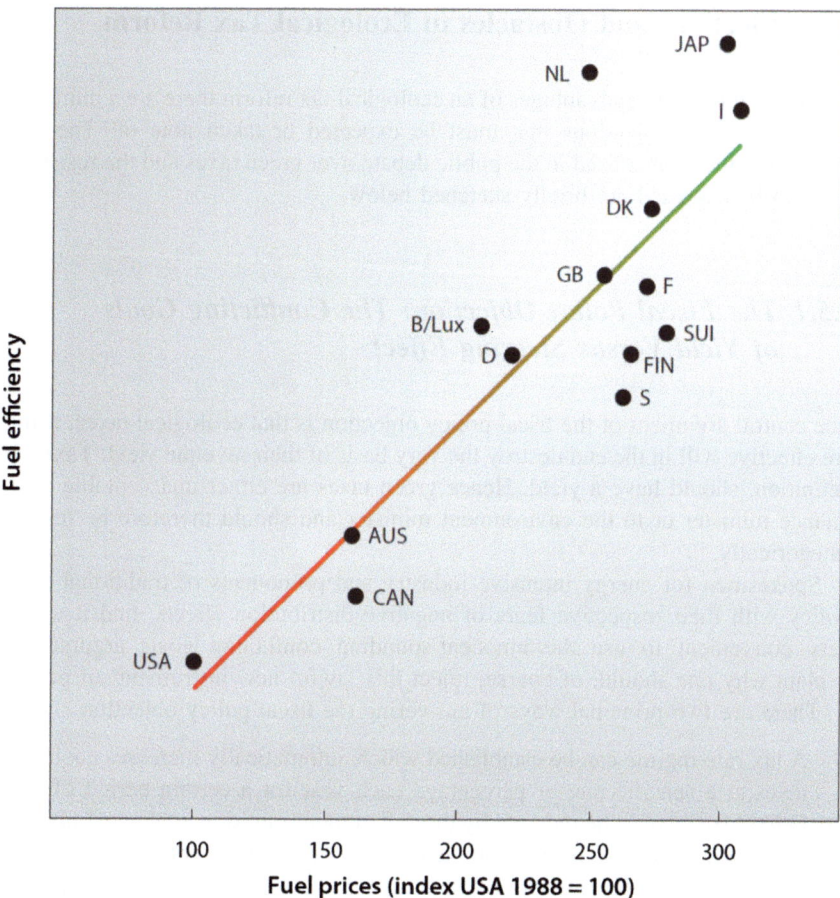

Fig. 9.2 Fuel prices of 1988 and 'macroeconomic fuel efficiency' in the most important OECD countries

inhabitants of countries of similar economic performance are in their use of fuel. Figure 9.2 shows the relationship between fuel efficiency and price for the year 1988:

The higher the macroeconomic fuel efficiency, the lower the specific fuel consumption. The correlation coefficient in Fig. 10.3 of $r = 0.85$ is not quite as high as in the previous graph. However, as opposed to the simple illustration in the first graph, even with high prices no negative consumption values appear here.

Naturally, Figs. 9.1 and 9.2 provide only a momentary view for 1988. For every year, slightly different but probably parallel lines would be obtained due to the fact the influencing factors keep changing. ….

9.5 Objections and Obstacles to Ecological Tax Reform

In spite of the evident advantages of an ecological tax reform there are a number of reservations and objections that must be expected or taken note of. The most important objections raised in the public debate over green taxes and the resistance that can be expected are briefly sketched below.

9.5.1 The Fiscal Policy Objection: The Conflicting Goals of Yield Versus Steering Effect

The central argument of the fiscal policy objection is that ecological taxes, if they are effective will in the end destroy the very basis of their revenue yield. Taxes, by definition, should have a yield. Hence green taxes are either unacceptable to the finance minister or to the environment minister and should therefore be rejected categorically.

Spokesmen for energy intensive industry and proponents of traditional social policy with their respective fears of negative distribution effects, find it always very convenient to use the innocent-sounding conflicting goals argument to explain why one should, of course, reject this 'awful new instrument' of policy

There are two principal ways of answering the fiscal policy objection.

(1) A tax rate regime can be established which automatically increases ecological taxes at a specific rate or percentage each year for a certain period of time. Other taxes would be reduced by the full amount raised in ecological taxes. In this way, revenues from ecological taxes grow sufficiently to please the Exchequer. At the same time, the steering effect would also grow steadily, which ought to please the environment minister. Obviously this mechanism does not remove the conflict in goals per se, but the conflict is no longer a reasonable ground for categorically rejecting the instrument of an ecological tax reform.

When after a certain time the steering effect becomes so strong that revenues are really shrinking, then the State will have to resort to more conventional taxes again (or to new ecological taxes).

(2) A similar means of overcoming the problem of conflicting goals might be to establish a certain tax revenue target and to adjust tax rates accordingly every year. This procedure likewise satisfies the Exchequer's needs. But it has the disadvantage of making energy and material prices unpredictable.

There is a historical example of an ecological tax in which a fixed revenue target was defined: the Japanese SO_2 tax. Revenue from the tax had to be maintained at an essentially constant level since it was earmarked for a fund designed to compensate pollution victims (so it was actually a special charge, not a tax). Advances in the technical control of emissions then led to an unexpectedly rapid decline in SO_2 emissions. Consequently, the tax rate had to be drastically increased and soon reached a level 500 times higher than in the beginning! To be sure, the Japanese economy as a whole was never substantially damaged by this. Although there was a strong rise in electricity prices, the production of flue gas desulphurization systems was given a major boost, and the economy quickly learned to live with higher power costs.

9.5.2 The Social Policy Objection: 'Green Taxes Aren't Fair'

Ecological taxes, and particularly taxes on energy, have a tendency to hit the less well-off layers of society relatively harder than the more affluent ones. This is a characteristic of almost all indirect taxes, especially of value added tax. The reason for this is that these taxes raise the prices of the basic things of life, which make up a smaller percentage in the shopping basket of the more well to do. While energy costs in wealthy households comes to between 2.5 and 3.5 % of annual expenditure, with retirees, the unemployed and other less well-off people, it is often in the region of or above 5 %. Figure 9.3 shows the social impact of ecological taxation.

Hansmeyer/Schneider (1989) and German industry (e.g. Förster 1990) cheerfully make use of the social policy objection as a sort of respectable support for their general rejection of ecological taxes. But there are also numerous warning voices from traditional left-wing parties. To answer this objection it may be useful to do some arithmetic.

Let us assume that energy taxes are raised at a rate of 5 % per annum. Let us also assume that energy productivity, or macroeconomic energy efficiency is increasing at a rate of 3 % per annum. Then the average additional expenditure on energy (assuming no increase in use) would be 2 % per annum. If energy costs make up 5 % of expenses in less well-off households, then the added annual costs resulting from energy taxes would be only 0.1 % on average which is not a figure to worry about in any major way.

If there was furthermore a political consensus to reduce VAT pari passu with increased energy taxes, then all potentially remaining negative distributional effects would disappear because the distribution effects of VAT are very similar to that of an energy tax.

Moreover, expanding public transport and simultaneously increasing the cost of air travel and heavy automobiles would actually have a positive distribution effect. (This is even more valid, by the way, in developing countries in which owning a car is still very much an exceptional privilege. Also domestic use of commercial energy is still correlated with affluence in the developing countries.)

Nevertheless, for reasons of political psychology (rather than out of real concern for equity) it may be useful to offer the ecological tax reform proposal in a

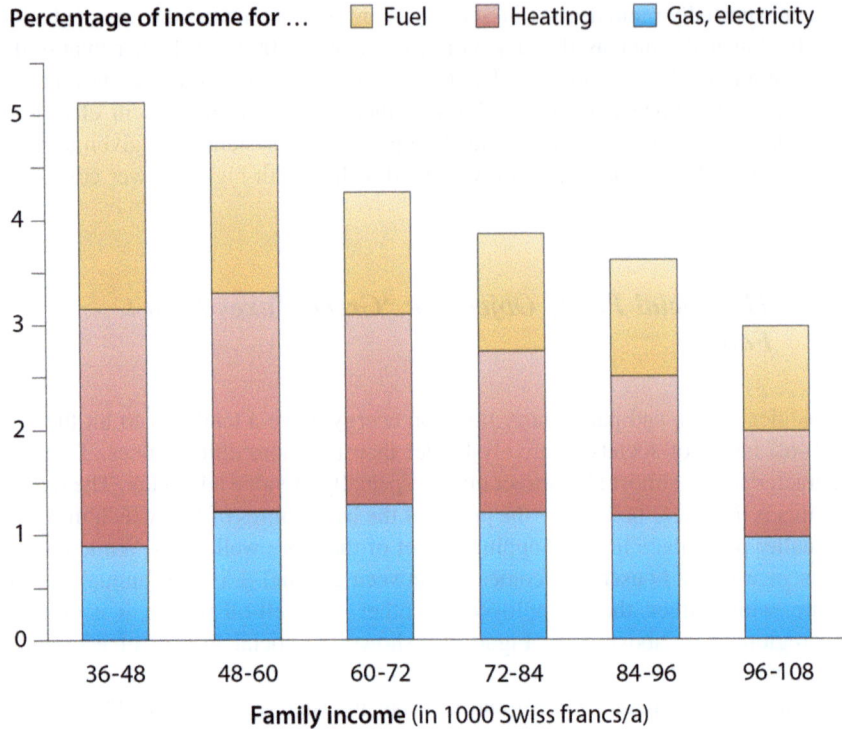

Fig. 9.3 Energy costs as a percentage of annual expenditure for Swiss households with various incomes (in Swiss francs) *Source* Mauch/Iten (1992) and Johnson et al. (1990)

package which includes some social measures to compensate the poorer strata of society for at least part of the added energy bill.

Finally, it must be said with regard to the social policy objection that the destruction of the environment is in itself 'regressive' in that it hits the poor harder than the affluent, who are far better able to escape its impact. If ecological tax reform is viewed as being an indispensable element in the ecological healing process, then it will be promoted precisely for reasons of social policy.

9.5.3 The Environmental Policy Objection: 'Revenue Should Be Purpose-Linked'

Environmental policy makers and the environmental interest groups tend to be preoccupied with the need for public expenditure for the sake of the environment. Comparing the budget of the environment minister with that of the social welfare or defence minister, one is tempted to believe that the environment minister's influence is far too small. Given this background, it is also tempting to believe that clearly all revenues from ecologically motivated taxes should flow directly into the coffers of the environment ministry and should be earmarked for urgent environmental tasks. Public acceptance, too, tends to be higher for purpose-linked charges than for taxes simply filling the coffers of a greedy, anonymous state. In terms of the words we use, that would prompt us to call them not ecological taxes, but special charges.

Naturally, it would be desirable both for the environment and for the environment minister if the proceeds from ecological taxes were placed directly at his or her disposal. But, as has been argued earlier, only an ecological tax reform has a chance to reach the magnitude of 5–10 % of GNP, while earmarked charges are restricted for economic reasons to very much smaller amounts.

Moreover, the legislature is always free to decide on spending priorities for non-earmarked revenues.

9.5.4 The Economic Policy Objection: 'Ecological Taxes Are a Burden on the Economy'

If ecological taxes are viewed merely as a means of increasing the cost of environmentally relevant production factors, one would fear that such taxes would damage the competitiveness of the economy. If, however, revenue neutrality is strictly observed, the damage would be counterbalanced by equivalent additional competitive advantages for those sectors that pay less taxes because of the reform.

Even so, some authors including Bergmann and Ewringmann argue that even a revenue-neutral ecological tax reform places a burden on the economy, in that it leads to avoidance investments that would normally not have been made. If the money devoted to such efforts is reducing funds otherwise available for investments in expansion and innovation, the reform might indeed work as a dead weight on the economy and competitiveness problems could theoretically ensue. But, in fairness, one should also ask how much money today is already being spent on avoidance investments against excessive labour costs, and how much of that is made profitable only due to the high taxes on human labour and how much society then has to pay for additional unemployment.

A careful, critical analysis should be made to find out whether or not an ecological tax reform works as a burden on the economy and how the burden—if any—can be avoided by appropriately adjusting the taxes. Such analysis cannot be done in a few pages. A slowly progressing, revenue-neutral ecological tax reform with provisions to prevent the ruin of any large-scale sectoral investment should be a relatively safe bet. And to make the comparison fair, it should be assessed against other measures that might otherwise be resorted to in order to slow down and ultimately reverse the dynamics of greenhouse gas emissions and to combat other global ecological disasters. Much will depend on the appropriateness of the price signal. The process of change should move forward at a pace which does not overtax an economy's innovative strength and readiness to invest. Furthermore, the greatest possible degree of international harmonization should be sought as we will argue later in this chapter.

Positive effects could come from two different corners: (1) if gross labour costs are reduced, positive employment effects could ensue which would relieve public budgets of an unemployment welfare payments burden and spur the economy further; (2) if the ecological tax reform creates a new and more reliable sense of direction for investments, the climate for investments may improve and the overall balance of spending may be shifted from consumption to investment. What should not be denied is the fact that there are going to be losers. Among these will be the following sectors—mining (abroad as well as at home), power plant construction, truck manufacturers, bulk chemicals, cement, and metal smelting plants (especially aluminium smelting from bauxite) Some sectors will find it impossible to hold out in their present form, and would have to emigrate. Aluminium companies would have to boost the recycling rate and expand their business with the aluminium service sector (which helps to save energy in other ways), and would in general have to diversify. But there is no pressing ecological reason why the losers should not be permitted to allow their existing plant to completely depreciate. At the same time, however, there is also no economic reason to keep on investing in ecologically unsustainable ways of production.

9.5.5 The Voter Objection: 'Ecological Taxes Aren't Popular'

The most important political task confronting those advocating an ecological tax reform is public information. It must be very clearly stated that ecological tax reform is not being introduced to finance environmental pipedreams. On the other hand, where there are already devastating budget deficits such as in the United States, Italy and Germany, it would not be wrong to use a portion of the revenue that results from green taxes for the purpose of reducing the deficit for a limited period. In this way—highly desirable both from a social and ecological standpoint—the trend toward higher interest rates could be weakened or reversed.

The hypothesis that an ecological tax reform would increase rather than decrease the amount of wealth available for distribution must be demonstrated in the most concrete and credible terms possible. Also, the failure of existing pollution control policies to address the global environmental problems effectively should be openly admitted, thus creating a desire on the part of the general public to do better. And the term 'ecological tax' should consistently be replaced by 'ecological tax reform'.

Above all, a cross-party consensus on ecological tax reform must be brought into being well in advance of any forthcoming important elections, thus preventing the reform from becoming a source of political dispute between the parties during the campaign. It should be clear to all responsible political parties that any idea of a significant, albeit slow pace for introducing an ecological tax reform must hinge on a broad political consensus which must not be put in question at every election. The consensus should reach the kind of stability enjoyed today by, for example, the necessity for a good educational system or social security.

9.5.6 The Effectiveness Objection: 'There are Instruments Which Work Faster and with Greater Precision Than Ecological Taxes'

Particularly in environmentalist circles, simply increasing the cost of environmentally harmful behaviour through taxation is often viewed as immoral: "The rich can buy their way out; it'll just be the poor who are forced to behave ecologically." For impatient environmentalists, clear-cut bans, ambitious standards, strict liability regulations, and strict environmental impact assessments in any planning process are considerably more appealing. In these circles, the ways in which costs ensuing from ecological taxes are passed on in higher prices is considered to be immoral, too, and is frequently cited as an argument against the effectiveness of ecological taxes.

Naturally, there are environmental challenges and tasks for which ecological taxes are not the answer. The ban on DDT and the impending prohibition of CFC's are certainly to be preferred over a mere tax-induced increase in the cost of the relevant substances (although in the case of those CFC's which are extremely useful substances for certain purposes, a charge-refund system might actually work faster and more efficiently than a ban which is not easily accepted by and enforced in countries like India or China). When there is immediate danger or when there is an acute health hazard, quick and drastic measures are needed. Nature conservancy and the protection of biodiversity, as well as the licensing of potentially hazardous industrial facilities, require conventional command and control measures including tougher liability laws, appropriate environmental impact studies, free access to environmental information plus some of the more conventional economic instruments.

However, economists like Bonus (1984) assert that economic instruments are likely to be twice as efficient as comparable regulatory measures. This is particularly relevant in view of the regulatory camps hopes that fuel efficiency in automobiles and heating efficiency in buildings can be brought about through obligatory efficiency standards. The result of this is sure to be far removed from the optimum path of transformation we have argued is necessary and possible, given an ecological tax reform.

Once again, it should be emphasized that the enormous effectiveness one can expect from an ecological tax reform has much to do with the long-term reliability, and hence predictability, of the process.

9.5.7 The Harmonization Objection: 'Ecological Taxes Cannot Be Harmonized Within the EC'

If one assumes that every environmental policy measure is a burden on the economy and will necessarily cause competitiveness problems at home, one would immediately wonder whether environmental laws are truly susceptible to being internationally harmonized. If the measures hold out a promise of working in an economically advantageous manner, however, the process of harmonization should be much easier. This is an argument for ecological taxes as an instrument of EC environmental policy or in the case of the USA, at federal level. A smaller harmonization problem may arise in relation to specific taxes. Article 99 of the EEC Treaty, for example, asks for harmonization of indirect taxes, thereby theoretically excluding the introduction at national level of green taxes before an EC-wide regime is established. But in practice there is scope for at least a certain individuality on the part of member countries, as has been allowed in the case of VAT.

Moreover, it was the Commission itself which took the first major initiative within the Community for introducing ecological taxes. In their proposal the Commission suggests, among a few other measures in the context of the fight against the accelerating greenhouse effect, the introduction in seven steps of fossil fuel taxes up to the level of $10 per barrel oil equivalent and half that fiscal burden for nuclear energy. The Commission also proposes strict budget neutrality and would allow some energy-intensive sectors to receive time-limited tax exemptions.

It may be worth mentioning in the EC context that an ecological tax reform should be highly attractive to those very countries whose poverty would normally be an obstacle to harmonization. This is because an environmental policy with relatively little in the way of administrative expenses and with the prospect of economic gain would, for such countries—Portugal or Greece spring to mind—presumably be more attractive than the present policy of imposing maximum permissible levels of pollution.

Along with all these objections that have been analysed above, and which should be taken seriously, there are a host of 9-day wonder objections which are essentially the result of misunderstandings. For instance, the occasional claim that ecological taxes are a form of 'strangle taxation' and are unconstitutional to boot, or the fear that ecological taxes are meant to replace existing regulatory measures, thus allowing any polluter who promptly pays his taxes to go on spewing filth into the environment. Or the fear that the tax bureaucracy could balloon enormously—which is a failure to recognize the fact that an ecological tax can and should be conceived of as simplifying and facilitating both the tax system and environmental policy.

9.5.8 Objections from the Losers

Along with political and scientific objections, outright resistance must of course also be reckoned with. This will come primarily from those who feel that they will be materially worse off as a result of ecological taxes. Broadly speaking, these include all those who consume natural resources to an above average extent or otherwise place an over-proportional burden on, or even destroy, nature. The relevant branches of the economy were indicated earlier in this chapter.

Losers can also mean specific population groups. First and foremost, motorists, but also frequent flyers, owners of detached homes, luxury hotel guests, big meat-eaters, and many others.

The organized representatives of labour, the retired and the unemployed could also offer political resistance if they lacked trust in the intended social neutrality of ecological tax reform, or sensed the chance of a redistribution in their favour if their negotiating stance were sufficiently tough.

Without a broad mobilization of winners and environmentalists, and without socially and economically acceptable arrangements, a broad and durable political consensus on ecological tax reform cannot be expected.

9.6 What Makes Ecological Tax Reform Attractive for the Business World?

Hardly any major political change is realistically conceivable these days that is not explicitly welcomed or at least tolerated by the private sector. Therefore it is important to assess the likely reactions of the business community to the idea of an ecological tax reform, and in terms more explicit than the general answers given elsewhere to the economic policy objection.

Ecological tax reform should significantly improve resource efficiency both on the micro- and macro-economic levels. Any investment that leads to output being maintained but with a substantially smaller input of resources (provided such gains in efficiency only require a reasonable level of investment expenditure) must be seen cumulatively leading to a macroeconomic advantage. Disposable wealth as a result is bound to increase rather than decrease. In the wake of an ecological tax reform. A steering effect which brings about an enhanced quality of life and a reduction in external costs in, for instance, the areas of medical care or pollution control ought to increase the amount of net wealth that can be distributed, even if there were no additional growth in gross national product.

For the business world, a long term strategy of a revenue-neutral and gradual ecological tax reform could become a highly attractive political goal. Clearly, the business world—like anyone else—is not going to favour an abandonment of other, more familiar policy elements But there are a number of reasons why business could come out in favour of such a long-tax reform term strategy. Reviewing these reasons is an appropriate way of bringing this short book to a conclusion.

(1) For business, long-term predictability is of first-rate importance. Were one to compare the concept of a quantitatively predictable tax reform running over a period of two to four decades with the leaps and bounds of current environmental policy and the incalculability and unpredictability of, say, the market for crude oil (on which the modern industrial economy will necessarily become increasingly dependent, the longer resource saving strategies are put off), an ecological tax reform ought to result in significant gains due to its enabling the private sector to avoid numerous false investment trails as well as due to its creation of an environment conducive to a stable and continuous process of innovation.

(2) For the economy—just as much as for the Earth's population—an effective strategy for reducing or even solving the global environmental problems that threaten (see Chap. 1) is something which is highly desirable. A positive commitment on the part of industry to such a strategy (rather than a merely grudging toleration of it) would, moreover, contribute positively to the building of a new image of industry and to the emergence of a new fundamental consensus in our societies. For its part, the formation of a consensus will have a further stabilising effect and so make possible more reliable predictions and again contribute to the safety of investments. A greatly to be desired consensus on energy policy would at last be within reach.

(3) If increased efficiency overall in the handling of scarce resources represents an economic gain, industry will also benefit from it. Likewise, it would by no means be the last to benefit from the economic gains which could result from a reduction in existing taxes on business profits, value added and gainful employment. And if private households begin to make energy-, water- and material-saving investments for their homes, in preference to forms of short-lived consumption like on frequent long distance travel, it could lead to an increase in demand for many industrial products. A whole new industrial growth sector might emerge.

Industry also plays a role as a customer of the capital goods sector and of consultancy and will increase its demand for innovative goods and services.

(4) The fact that an ecological tax reform can be put into force in developing countries in practice (and not merely on paper) and can perhaps be made politically acceptable in such countries, elevates this instrument well above the bureaucracy-intensive regulatory instruments as regards the potential for international harmonization Certainly harmonization in practice of pollution control standards between countries of different levels of development has proven exceedingly difficult.

This would make it easier for international companies to extend their philosophy of identical standards worldwide (which at present only applies to certain limited pollution control commitments) to the whole area of resource efficiency. Companies will have to watch out for criticism of their international operations with regard to the high rate of natural resource consumption (and pollution) involved in the early stages of production. These early stages of production typically are located in poor countries with often non-existent environmental controls.

(5) As soon as an ecological tax reform of the kind we propose shows visible signs of ecological success, it would become defensible to clear out the current regulatory thicket and significantly reduce the environmental bureaucracy presiding over it. On the other hand, if present rates of environmental destruction continue, public pressure could escalate sharply in favour of a further tightening of command and control instruments, including much more far-reaching liability laws which could turn almost every investment into an incalculable risk.

(6) A shifting of technological development in the direction of high sophistication low waste production again makes this instrument distinctly innovation-friendly. Countries and regions following this strategy are bound to obtain a competitive edge in modern production over those persisting with conventional methods of production and mere bureaucratic innovation policies This should also make for a highly attractive business climate for the high sophistication sectors of production.

(7) Finally, failure to adapt to the global environmental crisis at an early stage could become exceedingly expensive later on. But if we begin early, we still seem to have time for that slow, profitable no ruptures strategy.

Obviously, these reasons will appeal to different companies and industries to a very varying degree. And each one depends on an ecological tax reform being instituted slowly, reliably, on a long-term basis, and as an integral part of a credible policy mix.

The best guarantee of an economically sustainable reform process would be an initiative from the business community to join hands with sensible environmentalists and politicians in designing an effective and reliable strategy of sustainable development in North and South.

The Business Council for Sustainable Development appears to have taken such an initiative. In its final report, *Changing Course* (1992) a strong emphasis can be seen on prices reflecting ecological realities. This could well become the starting point for a joint strategy between business leaders and environmentalists.

References

Bonus, Holger, 1984: *Marktwirtschaftliche Konzepte im Umweltschutz* (Ulmer: Stuttgart).
Business Council for Sustainable Development, 1992: *Changing Course* (Geneva: BCSD).
Deudney, Daniel; Flavin, Christopher, 1983: *Renewable Energy: The Power to Choose* (New York–London: W. W. Norton).
Förster, Heike, 1990: *Ökosteuern als Instrument der Umweltpolitik? Darstellung und Kritik einiger Vorschläge* (Cologne: Deutscher Industrie-Verlag).

Hansmeyer, Karl-Heinrich; Schneider, Hans Karl, 1989: *Fortentwicklung der Umweltpolitik unter marktsteuernden Aspekten* (Bonn: Federal Environment Ministry).

IPCC (Intergovernmental Panel on Climate Change), 1990a: *The IPCC Scientific Assessment* (Cambridge: Cambridge University Press).

IPCC (Intergovernmental Panel on Climate Change), 1990b: *The IPCC Impacts Assessment* (Canberra: Australian Government Publishing Service).

IPCC (Intergovernmental Panel on Climate Change), 1990c: *The IPCC Responses Strategies* (Geneva: World Meteorological Organisation/United Nations Environmental Programme).

Johnson, Paul; McKay, Steve; Smith, Stephen, 1990: *The Distributional Consequences of Ecological Taxes* (London: Institute for Fiscal Studies).

Mauch, Samuel; Iten, Rolf, 1992: "Schrittweise Ökolosierung des schweizerischen Steuersystems", in: Mauch, Samuel; Iten, Rolf; Weizsäcker, Ernst U. von; Jesinghaus, Jochen (Eds.): *Ökologische Steuerreform* (Zurich: Rüegger Verlag).

O'Keefe, P; Pearsall, N.N. (Eds.), 1988: Renewable Energy Sources for the 21st Century (Bristol - Philadelphia: Adam Hilger).

Sterner, T, 1990: *The Pricing of and Demand for Gasoline* (Stockholm: Swedish Transport Research Board).

von Weizsäcker, Ernst U, 1991: "Sustainability—A Task for the North", in: *Journal of International Affairs*, 44: 421–432.

von Weizsäcker, Ernst Ulrich; Jochen, Jesinghaus, 1992: *Ecological Tax Reform. Policy Proposal for Sustainable Development* (London: Zed Books).

Chapter 10
Sustainable Energy Policies: Political Engineering of a Long Lasting Consensus

10.1 An Alliance of Long-Termers

At international meetings, the representatives of the energy business always strike me as genuine advocates of long term thinking. That's what I like about them. I simply assume that it should be possible to form an alliance with them, an alliance of long-termers. Other long-termers include educators, infrastructure planners, the wise men and women that exist in every society. The Club of Rome may also qualify as a part to the alliance of long-termers as all those memorable reports *to* the Club and one report *by* the Club of Rome had a distinct emphasis on long term consequences of what the world is currently doing. The first report to the Club, the famous *Limits to Growth* of 1972 laid the foundations to what later was to be known as *sustainable development*.[1]

Sustainability too is a symbol of long-termism. Since the publication in 1987 of the Brundtland Report the international debate has definitely seen a shift from short term crisis management to a more long term perspective. Sustainability has since become the most important yardstick for desirable development. *Yardstick* is actually not quite the right term because nobody has so far managed to measure sustainability.

Nevertheless we have a fairly certain judgement about processes that are *not* sustainable. To these belongs the burning of fossil fuels at the present rates, the present population growth or the rate at which new states are being formed in Eastern Europe.

Sustainable energy policies clearly require long term thinking. Energy investments tend to require an extended planning period, a couple of years for construction and a long period of time for the use of the investment. Stable economic conditions, predictable demand, secured fuel supplies and a stable societal acceptance are the ingredients for commercial success in the energy business.

On the other side of the intended alliance, the environmental camp, the aim is long term ecological stability. After the success story of pollution control (which

---

[1] This paper was presented at a workshop on "Sustainable Development and the Energy Industries" that was organized by Chatham House in London on 19 November 1993.

was not really a long term affair but rather one of crisis management) we environmentalists are turning our attention to such long term issues as the greenhouse effect, long term disposal of nuclear waste, genetic erosion, the ozone hole, population growth and changes in our life styles.

To begin with the latter, the life styles prevailing in the OECD countries are very far from being sustainable. A thousand Germans consume roughly ten times more energy and other limited resources per capita than a thousand people in less developed countries. If German consumption levels were the precondition for political stability, there would be no hope left for the global environment, because it is economically unfeasible and ecologically impossible to have German let alone North American consumption levels copied by 5.5 billion people.

People around the world have begun to realize that we are on a dangerously unsustainable trip. May I invite the energy community to agree with this statement? I also invite you to join hands with us environmentalists in seeking realistic long term strategies to bring us out of the present crisis.

10.2 Greenhouse Effect

The greenhouse effect may serve as an exemplary field of crisis suitable to study and recognize the magnitude of the problems we are facing.

The IPCC sees a reduction by some 60–80 % of greenhouse gas emissions as a necessary target to prevent a dangerous acceleration of global warming. What the climatic experts may still have underestimated is the threat of sudden sea level rises. The polar ice caps may be less stable than thus far assumed. Sea level rises in the past seem to have occurred in rather dramatic jumps of a couple of metres within decades (Tooley 1990).

Contrasting sharply with the needs established by the IPCC, some energy experts forecast increases of energy demand world-wide by some 50–70 % until 2020 which may be extrapolated to a doubling until 2040. Even the most aggressive build-up of nuclear power and renewable energy sources—at very high ecological and economic cost—would not be able to close the gap left between the IPCC exigency and GHG emissions associated with typical forecasts of energy demand.

No classical political measures are available to respond to this extraordinary challenge. CO_2 standards; forest protection; subsidies for afforestation, dry rice cultivation (to reduce methane emissions from rice paddies) or solar technologies; behavioural education; ecological management may all be desirable and necessary but are bound to miss the target by a wide margin. The climate challenge requires new thinking on a very fundamental level.

10.3 A New Direction for Technological Progress

We won't get it cheaper than by fundamentally redirecting technological progress. Which means that also values and cultural progress will have to be redirected?

Technological progress in the past has been characterized chiefly by the increase of labour productivity. Science and technology plus logistics and good management have enabled the fore-running countries to increase labour productivity by roughly a factor 20 over 150 years. Resource productivity was left behind in the process. It hardly increased at all, as can be seen from the fact that energy and material resource consumption grew almost in parallel with GDP in all industrialized countries. Only after 1973 a moderate decoupling occurred which was triggered by the oil price increases?

Now it's time to change the course. *Eco-efficiency* (Schmidheiny 1992) is the name of the new game. However, Schmidheiny's Business Council for Sustainable Development did not undertake to quantify the efficiency objectives. In terms of macroeconomic energy productivity, we must aim at *a factor of four* to close the gap indicated in the previous section. What the world needs we at the Wuppertal Institute are calling an efficiency revolution.

Quadrupling energy productivity, that 'revolutionary' objective, is not so outlandish as it may sound at first glance. It will be reached by a mere 3 % annual increase over some 45 years. And for many processes involving energy consumption a doubling of efficiency is possible even with existing technologies and without requiring any major changes of behaviour or infrastructure. If such deeper reaching changes are allowed, a quadrupling of macroeconomic energy productivity is conceivable using existing technologies.

The point of entry into rising energy productivity will be *Least Cost Planning* (LCP). If power plant permits are made dependent on the proof that there are no lower-cost alternatives available to fill the expected energy gap, it soon becomes more profitable for the utilities to subsidize energy efficiency at the consumer's end. Utilities are permitted to raise the price for the kilowatt-hour if the monthly bills of their customers are reduced. The profits from such efficiency investments will thus be shared between utilities and their customers. The capital savings for the utilities can be very substantial. Pacific Gas and Electric, one of the largest utility companies of the world has entirely scrapped its construction department and is making more profits than before. Similar developments are underway at Ontario Hydro, Canada's largest energy company.

However exciting this development is, it will not reach beyond relatively superficial savings to reduce excessive waste of energy. Truly new technologies of energy efficiency, new philosophies of low-energy food production, durability of goods and new infrastructures that help reduce the energy bill are not to be expected as a result from local LCP. The simple reason is that energy prices are too low to make major changes profitable.

10.4 Ecological Tax Reform

The straightforward way of driving technology development into the new direction should be to make it profitable. Energy prices should go up. Energy efficiency had its heyday during the 'energy crisis'. Prices can be raised by bureaucratic measures or by cutting subsidies or by establishing a regime of tradable energy permits or by raising taxes.

Tradable permits should be the preferred option for international negotiations. Based on assumptions of equity per capita emissions permits can theoretically be established. The South would clearly benefit from such allocation and would try to gain as much revenues as possible from selling permits to the North. Some concessional 'grandfathering' may be negotiated to reduce costs for the North. However, I remain rather sceptical if the North will really accept a regime that turns out to be *very* costly for Northern economies.

Anyway it is fairly difficult to imagine how a system of tradable permits could work at *domestic* markets. Who is there to measure and control carbon dioxide emissions from wood stoves or methane emissions from compost? Even for fossil fuels the establishment of a carbon dioxide market would involve considerable additional monitoring and control costs. Moreover, permit prices are likely to increase sharply in economic boom times, causing severe social problems with poor people who may not enjoy any benefits from the economic upswing.

By comparison, all such problems of administrative cost, unpredictability of prices and social equity can be greatly reduced by a strategy of direct pricing of carbon dioxide emissions or of energy. I am outlining such a strategy which may be labelled 'Ecological tax reform'.

The strategy would begin by reducing subsidies on basic commodities. Why should coal mining, nuclear energy or agricultural products be subsidized by taxpayer's money? Even 'subsidies' (i.e. public spending) for roads or airports are ultimately irrational. He who wants transport ought to pay the full price. It is an economic fallacy to believe that transport *per se* is a good thing for the economy. It is only a good thing for volume and turnover of the economy. But, then, even traffic accidents add to 'economic' turnover without making anybody happier.

Cutting subsidies is not enough. Externalities should also be accounted for. For this, taxes may be levied on non-renewable sources of energy, on primary raw materials, on water consumption, on certain chemicals such as chlorine or metals or on certain types of land use. (I remain sceptical against taxes on polluting emissions or on waste owing to evasion and control problems chiefly in less developed countries).

Externalities need not be calculated in detail. Crude estimates suffice. Taxes have the advantage of not requiring erudite justifications via the computation of externalities. Income taxes or VAT were never justified this way. But, according to Arthur Cecil Pigou, the economy as a whole would benefit if taxes would *more or less* reflect the societal costs of energy and other basic commodities. Studies such as the one on energy externalities commissioned by the German ministry of economic affairs (Prognos 1992) seem to indicate that externalities lie in the vicinity of 10 % of GDP.

To avoid any expansion of public spending, other taxes, charges and levies should be reduced by equivalent amounts. Notably one should reduce the fiscal and parafiscal burdens on human labour thus making labour more affordable again for employers.

The cutting of subsidies plus ecological tax reform would gradually make the increase of resource productivity more profitable and the dismissal of workers less. Repetto et al. (1992) quoting Ballard/Medema (1992) seem to have proven Pigou's assumption and state that "the total possible gain from shifting to environmental charges could easily be $0.45–$0.80 per dollar of tax shifted from 'goods' to 'bads'—with no loss of revenues".

The greening of subsidies and ecological tax reform may be designed such that end user prices for ecologically problematic factors increase steadily and predictably by constant factors over many decades (von Weizsäcker/Jesinghaus 1992). A reasonable first approximation may be a price increase by 5 % annually which would lead to a doubling in 14 years, a quadrupling in 28 years and an octupling in 42 years. This is a very strong signal which can easily change the course of technological progress.

And yet, owing in part to expected productivity gains (estimated conservatively to 3 % annually), the price signal would be extremely tame. It would be 2 % (5 − 3 %) annually for a production factor weighing on average less than 4 % of total production costs, so that the remaining cost differential would be only 0.08 % annually. Even this almost imperceptible cost differential would—on average—be outbalanced by labour cost reductions e.g. from reduced social security payments. Hence firms would on average even financially benefit from the reform.

To be sure there would be losers. Aluminium smelters (from bauxite), steel, bulk chemicals, cement, pulp and paper and a few others would run into difficulties even if the price signal remains as tame as suggested. To avoid additional unemployment it may be considered to give such firms a temporary tax relief until their existing capital stock is written off. But investments into the same old-fashioned production should be prevented.

If the WRI researchers are right, one should expect far more winners than losers. I am convinced that energy companies can comparatively easily restructure in a way making them winners as well. The centrepiece of their future business would be energy services: selling packages of energy management for homes, factories, hospitals or office buildings. There are hardly any better experts conceivable in this field than the present energy companies. Petrol companies (which are already now diversifying into many fields) could move into the mobility market, competing there with logistics firms, car manufacturers and car rental services. They may specialize in fuel efficiency for the car and lorry fleets of larger firms.

At the political arena, the main problems of the tax strategy are predictability and international harmonization. But if there is a societal consensus that the scheme is beneficial for the economy, not only for the environment, it should not be too difficult to agree on it among all major political parties thus keeping it out of election campaigns. International harmonization will be much easier than is the case for classical environmental policy measures because the latter invariably involve extra costs with no immediate economic benefits. Also, the cutting of subsidies is recommended to less developed countries anyway, and simple resource taxes are far easier to collect than personal taxes as everybody will admit who has ever lived in a developing country.

In addition to the theoretical macroeconomic gains which the WRI team (Repetto et al. 1992) expects from ecological tax reform there may be still more attractive prospects involved in the idea.

If business leaders and capital owners see scope for a new and reliable road of technological progress they may feel encouraged to pour money into the new trend. There is hardly anything more stimulating for the business world than a steady and predictable trend. Once the consensus is strong enough to persuade the pioneers, the process can soon become self-enforcing and may lead to a new 'Kondratiev' cycle.

Is it not an extremely tempting idea to use the high certainty of the environmental crisis as a basis for a high certainty of a new technological trend? Is it not, therefore, a good idea to put the shift of technological progress on the agendas of the G 7, of OECD or of the relevant UN bodies?

10.5 Transport Costs and International Trade

Implementing the above strategy of 'making prices tell the ecological truth' could have an important side effect. High energy and resource prices also mean higher transport prices. The transport sector has been a favourite target for state subsidies. As long as labour productivity increases was the main objective for technological progress, free infrastructure use and low taxation on transport seemed quite rational. But once resource productivity becomes an equivalent objective, excessive transport will be discovered to be one of the main sources of wasteful resource use.

International trade is greatly benefitting from the infrastructures which once were paid by taxpayers. Development aid has for a long time favoured the construction of ports, roads, railways and airports. This made primary commodities much more accessible which via international competition have contributed to increasing commodity supplies an gradually falling prices for them.

Now it is the time to recognize that subsidized transport costs can actually be destructive not only to the environment but also to the economy. What is the use of spending scarce public funds for making foreign goods artificially cheaper than they would naturally be? And what is the use for developing countries to ever further specialize in commodity exports which hardly contribute to higher development? Higher transport prices will render the criss-crossing across the oceans of low-value goods unprofitable and will move international trade towards high-value goods.

Making transport prices tell the ecological truth is perhaps the only legitimate 'protectionism' and the least bureaucratic and discriminatory too.

A new equilibrium will have to emerge between economies of scale and resource efficiency. The 'depth' of manufacture will increase again as the supply of parts from very distant places becomes an *apparent* absurdity. (It's a macroeconomic absurdity *now* in many cases but appears rational on the microeconomic levels).

Rising transport prices may give some relief to industries and farms suffering from ruinous competition from abroad. Local markets are likely to re-gain some economic importance which may also be good for social cohesion.

For commodity exporting countries including the OPEC and for certain industrial branches such as bulk chemicals, cement, metals and long distance hauling the shift may at first look unpleasant. However they cannot count with a continuation of the present development forever. For them a slow and well-planned phase-out of unsustainable processes may be the best they can get. Following the example of the North American utilities they may even with their specific knowledge become active—and profitable—players in the transformation process. The may become the first suppliers of commodity-related *services* and make better profits from using up less of those commodities.

At the international level, some compensation measures for commodity exporting countries are well conceivable. As I am assuming that the Northern economies would greatly benefit from a strategy of shifting towards resource productivity I don't see why the North cannot afford some additional official development aid directed to the necessary restructuring of former mono structures of commodity exporters.

References

Ballard, Charles L.; Medema, Steven G., 1992: *The Marginal Efficiency Effects of Taxes and Subsidies in the Presence of Externalities: A Computational General Equilibrium Approach* (East Lansing: Michigan State University, Department of Economics).
Prognos, 1992: Externkosten der Energieerzeugung. Gutachten für den Bundesminister für Wirtschaft (Bonn: BMW).
Repetto, Robert; Dower, Roger C.; Jenkins, Robin; Geoghean, Jacqueline, 1992: *Green Fees: How a Tax Shift Can Work for the Environment and the Economy* (Washington, D.C.: World Resources Institute).
Schmidheiny, Stephan; Business Council for Sustainable Development, 1992: *Changing Course. A Global Business Perspective on Development and the Environment* (Cambridge, Mass.: MIT Press).
Tooley, Michael, 1990: "The Impacts of Sea-Level Changes—Past and Future", in: Gerling Globale Reinsurance Company (Ed.): *Changing Weather Patterns* (Cologne: Gerling Globale Reinsurance Company).
Weizsäcker, Ernst Ulrich von; Jesinghaus, Jochen, 1992: *Ecological Tax Reform. Policy Proposal for Sustainable Development* (London: Zed Books).
Weizsäcker, Ernst Ulrich von, 1994: *Earth Politics* (London: Zed Books).

Chapter 11
Factor Four: Doubling Wealth—Halving Resource Use: A New Report to the Club of Rome

Ernst Ulrich von Weizsäcker, Amory B. Lovins and L. Hunter Lovins

11.1 Preface

This is an ambitious book which seeks to redirect technological progress.[1] Aggressive increases in labour productivity constitute a rather questionable programme at a time when more than 800 million people are out of work. At the same time, scarce natural resources are squandered. If resource productivity were increased by a factor of four, the world could enjoy twice the wealth that is currently available, whilst simultaneously halving the stress placed on our natural environment. We believe that we can demonstrate that this quadrupling of resource productivity is technically feasible and would produce massive macro-economic gains, i.e. make individuals, firms and all of society better off.

As we outline in this ground-breaking programme, we are taking up the concerns expressed in the early 1970s by the Club of Rome, which shook the world with its report The Limits to Growth (Meadows et al. 1972). This time, however, we give an optimistic answer to these concerns. We will demonstrate that equilibrium scenarios are available. Factor Four, we believe, can put the Earth back into balance (to adapt a metaphor from Al Gore's challenging bestseller (Gore 1992)).

We wish to thank the Club of Rome for a steady and growing interest in our project. A Club of Rome ad hoc seminar was organized in March 1995 in Bonn, co-sponsored by the Friedrich Ebert Foundation and the German Environment Foundation, to discuss the manuscript of the book. As a result, much of the text was rewritten and sent to members of the Club's Executive Committee, which in

---

[1] This text was first published in the book coauthored with Lovins, Amory; Lovins, Hunter: *Factor Four. Doubling Wealth, Halving Resource Use* (London: Earthscan, 1997). Permission to republish this text was granted by Lizzy Yates, Permissions Administrator, Taylor & Francis Royalties Department, Cheriton House, North Way, Andover, Hampshire, UK on 19 February 2014.

June 1995 accepted the book as a Report to the Club. We are honoured that the Club's President provided a Foreword to this edition.

The manuscript was originally written in various versions of English—half of it by a native German speaker, half by two Americans. ...Amory B Lovins and L Hunter Lovins published a complementary book, with Paul Hawken, framed for the US ...context and primarily for a business audience (Paul Hawken 1998).

We are greatly indebted to all those who were engaged in the discussion of this book even before it became available in what we hope is a more faithful version of the English language. Hundreds of people have been involved in the making of this book ... Without the pioneering work of Herman Daly, Dana and Dennis Meadows, Paul Hawken, Hazel Henderson, Bill McDonough and David Orr, it would have been almost impossible to conceive of a book with this scope.

We also thank the sponsors of the Bonn meeting and the government of North Rhine Westphalia for the basic grant given to the Wuppertal Institute for Climate, Environment and Energy, as part of the North Rhine Westphalian Science Centre, with the injunction to investigate and put into practice the principles of this book.

Ernst von Weizsäcker, Amory B. Lovins, L. Hunter Lovins
January 1997

11.2 Introduction: More for Less

11.2.1 Exciting Prospects for Progress

'Factor Four', in a nutshell, means that resource productivity can—and should—grow fourfold. The amount of wealth extracted from one unit of natural resources can quadruple. Thus we can live twice as well—yet use half as much.

That message is both novel and simple.

It is novel because it heralds nothing less than a new direction of technological 'progress'. In the past, progress was the increase of labour productivity. We feel that *resource productivity* is equally important and should now be pursued as the highest priority.

Our message is simple, offering a rough quantitative formula. This book describes technologies representing a quadrupling or more of resource productivity. Progress, as we have known at least since the Earth Summit at Rio de Janeiro, must meet the criterion of sustainability. *Factor four* progress does.

The message is also *exciting*. It says that some aspects of that efficiency revolution are available at *negative cost*; that is, profitably. Much more can be made profitable. Countries engaging in the efficiency revolution become stronger, not weaker, in terms of international competitiveness.

That is not only true for industrialised countries of the North. It is even more valid for China, India, Mexico or Egypt—countries that have a great supply of inexpensive labour but are short of energy. Why should they learn from the US and from Europe how to waste energy and materials? Their journey to prosperity will be smoother, swifter and safer if they make the efficiency revolution the centrepiece of their technological progress.

The efficiency revolution is bound to become a global trend. As is always the case with new opportunities, those who pioneer the trend will reap the greatest rewards.

11.2.2 Moral and Material Reasons

Changing the direction of progress is not something a book can do. It has to be done by people—consumers and voters, managers and engineers, politicians and communicators. People don't change their habits unless they have good reasons for doing so. Motivation needs to be experienced as compelling and urgent by a critical mass of people, otherwise there won't be enough momentum to change the course of our civilisation.

The reasons for changing the direction of technological progress are both moral and material. We trust that most readers share our view that preserving physical support systems for humankind is a high moral priority. The ecological state of the world demands swift action as will be discussed in Part III. We avoid the language of doom and gloom, but we do present some disturbing ecological facts and trends. These should be established if we want to say something in quantitative terms about the necessary answers. We shall demonstrate that gaps as large as a factor of four are opening before us which need to be closed.

If these gaps aren't closed, the world may run into unprecedented troubles and disasters. Avoiding them may seem like a formidable task. Can such gigantic gaps be closed at all? This question leads us to the good news. The gaps can be closed. *Factor Four* is at the heart of the answer. Best of all, we shall discover very strong *material* reasons for changing the direction of technological progress.

Countries starting at once will reap major benefits. Countries that hesitate are likely to suffer formidable losses of their capital stock which will quickly become obsolete as resource efficiency trends take hold elsewhere.

11.3 Chapter 1: Twenty Examples of Revolutionising Energy Productivity

People used to talk about 'saving energy'. The phrase had a moralistic connotation. Father would admonish his children to switch off lights when leaving a room and never to let motors or appliances run when not needed. After all, besides costing money, wastefulness was a sin. When a demand for environmental protection became widespread, the reaction on the part of governments, electric power suppliers and some environmental leaders was not particularly imaginative: You (childish and demanding folks) can get as much environmental protection as you want if you are prepared to radically reduce your demands. The simplistic notion of saving energy by voluntarily making do with less allowed the leadership to avoid really grappling with the energy issue in a creative way.

In recent years, a new term has emerged: the rational use of energy. Use of this term enhances the speaker's reputation by suggesting an expertise in energy matters. So we may hesitate to reject this term, yet we are not happy with it either. It sounds so bureaucratic, complicated and defensive. It doesn't convey any enjoyment and is not straightforward in talking about the relationship between energy use and technological progress.

Technological progress is what this book is all about. Or, rather, it is about redirecting technological progress. Thus we prefer to speak of energy productivity.

Taken by itself, and depending on your circumstances, the term productivity can have positive or negative connotations. These mixed connotations spring from the disservice of economists who have narrowed down the term to mean only labour productivity. In the past, labour productivity was a good thing meaning prosperity. Today, labour productivity is inevitably associated with the threat of unemployment.

Energy productivity, on the other hand, is something everybody can greet with joy. Virtually nobody stands to lose by it.

This chapter is about increasing energy productivity by a factor of four. The terms energy savings or the rational use of energy are simply inadequate to conveying the appropriate sense of joyous attack on our prevalent technological dinosaurs. The concept of energy productivity is more equal to the task.

By using Factor Four as a standard, we appear to exclude much of the manufacturing world. Smelting aluminium from bauxite cannot, given the laws of thermodynamics, be made four times more energy efficient. The same holds for chlorine, cement, glass and some other basic materials. But we need not give up on the Factor Four potential of these materials. Aluminium and glass are superbly recyclable, and such recycling would save a lot of energy. For some end uses, some materials can be substituted for, with no damage to the manufacturing sector; or materials can be used more judiciously. On a life cycle basis a factor of four in energy productivity should be available for most end user services involving metals or glass.

In this book, however, we concentrate on examples with a straightforward potential of multiplying energy efficiency by four or more. We begin with an example that has an overwhelming importance for the worldUs energy balance.

(Twenty examples of a fourfold increase of energy productivity—or more—follow this introduction.)

11.4 Chapter 2: Twenty Examples of Revolutionising Material Productivity

Material resource productivity is not a familiar concept. We owe it mostly to Friedrich Schmidt-Bleek, the Director of the Wuppertal InstituteUs Division for Material Flows and Eco-Restructuring. Schmidt-Bleek has developed the Material Inputs per Service unit (MIPS) concept, a way to determine or estimate for any well-defined service the kilograms or tons of materials that have to be moved around somewhere in the world. For a given service, material inputs might include the tailings from a copper mine in Chile, plus the water and other materials used and moved in the manufacturing process in Mexico, plus the packaging done in Chicago, plus some materials used and moved in the final sales process. Chapter 9 explains the MIPS concept in more detail.

Material productivity, then, refers to the reduction of MIPS. Longevity of products obviously helps to increase material productivity, where the services rendered are not significantly compromised as time passes. Think of old furniture which may actually gain in value over time. On the other hand, longevity stands in conflict with modernity, fashion, and technical performance (including efficiency). Material productivity is a wider notion than durability; it refers to the product life cycle 'from cradle to grave.'

Schmidt-Bleek (1994) feels that a factor of four in MIPS reduction will not be enough. A factor of ten (see Sect. 9.3) is what he believes is necessary for the OECD countries. We hope our friend will excuse our pusillanimity when we nevertheless dare to include many examples showing a mere factor of four. Let's agree to call it a worthy beginning.

The Product Life Institute in Geneva, directed by Walter Stahel, has developed strategies for optimising resource efficiency. The context is the 'Service Economy' in which it's the end-user service that counts (Giarini and Stahel 1989/1993). The following elements may be of use for such a strategy:

- Leasing instead of selling, wherein the manufacturerUs interest lies with durability;
- Extended product liability, which could induce manufacturers to guarantee low-pollution use and easy re-use or disposal;
- Joint ownership or use (e.g. of cars or appliances), which would require less products for the same amount of services;
- Re-manufacturing preserving the stable frame of a product after use and replacing only worn-out parts;
- Product-design optimised for durability, re-manufacturing and recycling.

Clearly, these elements indicate a complex, multi-task strategy. The material flows resulting from performing a task depends on how much of the task we do; how efficiently we use the materials required to do the task; how much ore has to be extracted and processed to obtain the materials needed; how far everything has to be shipped; and how many movements of material were involved in earlier years to create the infrastructure, the factories and the distribution networks. Material efficiency at every stage of the process determines, in a cumulative fashion, how much we can do with how little.

If everyone in our society is to have one widget, how many widgets do we need to make each year? Just enough to keep up with the number that break, wear out, or are sent away, plus however many we need to keep up with net population growth. The key variable is clearly how long the widgets last. To have something to drink out of, we need a lot fewer ceramic mugs than paper or Styrofoam cups, because the ceramic lasts almost forever (unless we drop it), while the 'consumer ephemerals' are used just once or twice and then thrown away before they fall apart. And if we make the ceramic mug unbreakable, it lasts long enough to hand on to our great-grandchildren. Once enough such unbreakable mugs are made for everybody to have one (or enough), very few need be made each year to keep everyone perpetually supplied.

(Twenty examples of a fourfold increase of material resource productivity—or more—follow this introduction.)

11.5 Chapter 3: Ten Examples of Revolutionising Transport Productivity

We are devoting a separate chapter to transport productivity, although each transport of goods or people involves both energy and material consumption. On the other hand, not all environmental impacts of transport relate to energy and materials. Habitat destruction (by roads), noise, mass tourism and ever increasing access to natural resources involve more than resource considerations, and so need to be considered separately. Certainly, transport-environment conflicts have a prominence of their own. Any efficiency gains in transportation will therefore be greatly welcomed outside all resource considerations. Moreover, the description of ways and means to quadruple transport productivity will yield insights about a new civilisation that we will have to develop beyond the efficiency revolution.

11.5.1 Videoconferences

Data highways have become one of the most powerful symbols of technological progress. Al Gore's 1992 best-seller, *Earth in the Balance* Gore (1992), has helped to build a broad public awareness of the important role that data highways can play in harmonising ecological and prosperity objectives. In this subchapter we are exploring in preliminary quantitative terms the potential of long-distance electronic communication to help multiply resource productivity.

Not surprisingly, we have found that the potential is far bigger than a factor of four. We did a rough estimate for two different cases:

- substituting electronic mail for posted letters; and
- substituting a video-conference for a business meeting.

The Rocky Mountain Institute (RMI) has been one of the pioneers in the systematic use of long-distance data communications. At RMI, video-conference apparatus that digitally compresses full-duplex colour and audio so it can be sent over a data-quality telephone line joins with e-mail, modem text and graphics transmissions, telephone, and fax to displace much travel. Soon after its installation in 1993, for example, the video-conference equipment enabled one of us to avoid 4 days' travel to and from Western Australia, at a few percent of the cost of the plane ticket, and with none of the inconvenience or fatigue. A major conference could be keynoted, complete with projected graphics, simply by sitting in a comfortable chair at RMI in front of the video-camera, pressing a few buttons to dial the apparatus at the other end and speaking normally.

Microchips programmed with Israeli data-compression algorithms sent images of what was moving (lips, eyebrows, etc.) but didn't keep re-sending parts of the image that didn't move (ears, background, etc.). The compressed signal went through a few copper wires to Basalt, Colorado; by optical fibre to Denver; by a series of satellites to Perth; by fibre again to Fremantle; by an improvised microwave link from the roof of the Telecom building to the roof of the conference hotel; by coaxial cable into the ballroom; through similar chips that spliced a high-quality, full-motion, full-colour picture seamlessly back together; into a video projector; and less than a quarter-second after a word was uttered in the mountains of Western Colorado, there it was, perfectly synchronised with digitally echo-cancelled sound, on the retinas and eardrums of the audience in Fremantle.

Returning to the standard question of this book, we may ask how much resources can be saved by telecommunication. Methodological questions arise over what to count and to compare. We opted for Schmidt-BleekUs 'MIPS'-method, the calculation of material inputs per service unit (cf. Chap. 9 and the introduction to Chap. 2). Hartmut Stiller of the Wuppertal Institute and Thomas Egner of the Ulm-based FAW (Forschungsinstitut für anwendungsorientierte Wissensverarbeitung) obtained the following provisional results.

For a trans-Atlantic business trip what has to be counted is the ecological rucksacks of the proportional air fuel consumption, of the aeroplanes, the stay in a business hotel and a few other items relating to the trip. As an estimate, the total ecological rucksack is about one ton. A 6 h video-conference, by contrast, may account for material inputs of rather less than 10 kg. This means that a half-day video conference could represent (with a wide margin of arbitrariness) a MIPS reduction factor of roughly a hundredfold.

Clearly, these results have to be used with a great deal of caution. Not all business trips can be adequately replaced by video-conferences. Larger conferences derive much of their value from coffee break chats, poster sessions, ad hoc discussions and delightful dinners, from side-programs and from the initiation and renewal of personal friendships. None of this is susceptible to electronic data transmission. Moreover, e-mail and video-conferencing create their own growing demand and can even incite participants to plan additional journeys they would not otherwise have thought of. So whatever the mathematical results are of the comparison between physical transportation and virtual or electronic transportation, the real world reduction of resource consumption may be quite modest, rather in the vicinity of a factor of four.

On the other hand, the huge reduction potential may legitimately play a role if policies are adopted to 'make transport prices tell the ecological truth'; i.e., making transportation significantly more expensive. In that case, many people will more readily forego certain trips, and take consolation in realising that much of the transport demand is, in reality, based not on need but on thoughtlessness.

The applications of video-conferencing are many and exciting. The marketing of special goods, even works of art, can be done via telecommunications. A popular chain of photocopying shops in the United States is rapidly installing equipment so that someone in any city can hold a video-conference with someone in another city without prior arrangements. Tens of thousands of private installations also exist. For example, we recently discussed some technical apparatus by video with a colleague by courtesy of an art gallery that normally uses the equipment to show objets d'art to prospective purchasers.

Perhaps one of the biggest growth markets for telecommunications is 'telecommuting', which normally involves data exchange and could involve video-conferencing. Many office tasks can be done—and are increasingly done—at a distance. Especially for families with small children, this represents a welcome opportunity to continue working without necessarily leaving home. To take another example, maintenance and repair does not always require the personal appearance of service experts. Much can be done via video-conferences.

(Nine more examples of a fourfold increase of transport productivity—or more—follow this introduction and the core of the first example—which was written mostly by Amory Lovins.)

11.6 Chapter 11: We May Have 50 Years Left to Close the Gaps

In this third part of our book we wanted to convey a sense of urgency for changing the present course of technological and civilisational development. Our intention is not to be doomsdayers, but to realistically assess the speed and momentum of the current trends, including growth rates in China and India, and investors' fascination with those growth phenomena.

Once infrastructures, urban developments and major capital investments worth billions of dollars are there, they will to a large extent determine the subsequent moves. Even readily available potentials of resource efficiency are likely to be ignored if creditors and owners want to see higher returns on their investments (and continue to believe that efficiency will raise rather than lower their costs).

In this book we are not emphasising points of no physical return from various aspects of ecological damage. Rather, the sense of urgency arises from fast approaching points of *economic* no return, after which it will be very costly to achieve the benefits of resource efficiency. But if we redirect investments and technology development *now*, we are likely to earn most of the expected benefits at negative cost.

Of course there *are* points of no physical return. Extinct species cannot be brought to life again, and major climatic changes will allow for a "return to normal" only in time spans of thousands of years. Assessing the scientific analyses on climatic change and other ecological menaces, we believe that the world may have some 50 years left to close the gaps.

This final chapter of Part III will be devoted to a more or less quantitative analysis of the chances of closing these gaps. Let us take the *Limits to Growth* study as our starting point.

11.6.1 Beyond the Limits? The Meadows May be Right

Donella and Dennis Meadows, together with Jørgen Randers and William Behrens, were the authors of the first major Report to the Club of Rome, *The Limits to Growth*, published in 1972. Some 9 million copies were sold in 29 languages, and the book changed the world. The global civilisation became aware of outer limits that had previously been ignored. Figure 11.1 shows the essential results of the model.

Not surprisingly, it didn't take long for the critics to discover flaws contained in the message, in certain details, and in the methodology. Poorer countries (and the poor living in rich countries) found it unfair that the rich should declare limits to growth at a time when the poor were just starting to experience economic growth. Resource specialists were able to show that mineral resources, including gas and oil, were far more abundant than the *Limits* were assuming. Many scientists, economists and politicians felt that *Limits* was too pessimistic in general, and John Maddox (1972) and his followers in the scientific and engineering communities, said that technological progress had so often produced unexpected answers to any given problem that it was nonsense to speak of rapidly approaching limits (Fig. 11.2).

Indeed, *The Limits to Growth* was based on a simplistic computer model and the results were also very simple. Some of the input data proved wrong. And technology can indeed do fabulous things. *Factor Four* actually wishes to be quoted as witness for this fact. On the other hand, the physical state of the environment *has* deteriorated to a large extent. The trends of growth have continued and have brought us much closer to certain limits, if perhaps others than those identified in the 1972 study.

State of the World

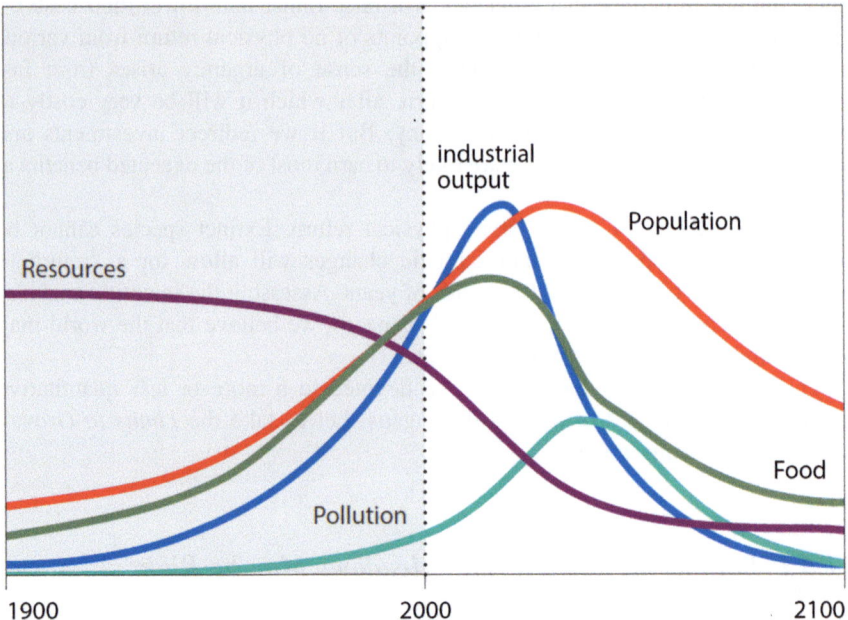

Fig. 11.1 The "standard run" from *The Limits to Growth*. Population and industry output grow until a combination of environmental and natural resource constraints eliminate the capacity of the capital sector to sustain investment. (From Meadows et al. 1992, p. 133). Meadows, Donella, Dennis Meadows and Jorgen Randers, 1992: *Beyond the Limits: Confronting Global Collapse, Envisioning a Sustainable Future* (Post Mills: Chelsea Green)

The Meadows and Jørgen Randers got together again 20 years later to work on an updating for their old book. But they discovered that they would have to write a new book altogether. Too much had changed, say the authors in the preface of their new book, *Beyond the Limits* (Meadows et al. 1992, p. xiv):

> As we compiled the numbers, reran the computer model and reflected on what we had learned over two decades, we realised that the passage of time and the continuation of many growth trends had brought the human society to a new position relative to its limits. In 1971 we concluded that the physical limits to human use of materials and energy were somewhere decades ahead. In 1991, when we looked again at the data, the computer model, and our own experience of the world, we realised that in spite of the world's improved technologies, the greater awareness, the stronger environmental policies, many resource and pollution flows had grown beyond sustainable limits.

For example, between 1970 and 1990, human population has increased from 3.6 to 5.3 thousand millions, automobiles from 250 to 560 million, natural gas consumption from 31 to 70 trillion cubic feet per year and electric generating capacity from 1.1 to 2.6 thousand million kilowatts. Whatever geologists may say about rich and undiscovered resources, consumption cannot continue to grow at this rate.

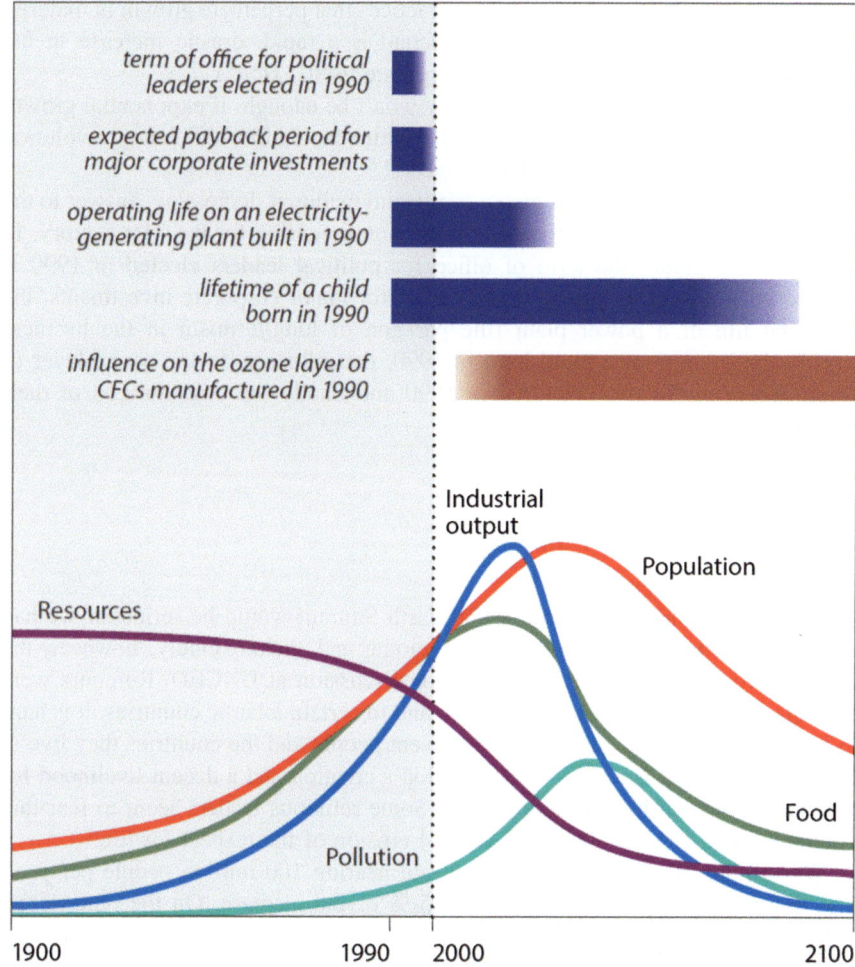

Fig. 11.2 Time horizons of politics, the corporate world, children, the environment and the system dynamics of the World3 model on which Fig. 11.1 were based (From Meadows et al. 1992, p. 235)

Even a stabilisation of consumption is not going to solve the problem. And many analysts say that it's not so much scarce resources but the absorptive capacity of the earth for all the pollutants and wastes that is limiting further growth of resource consumption.

We have come closer to the limits, say the Meadows: "Without significant reductions in material and energy flows, there will be in the coming decades an uncontrolled decline in per capita food output, energy use and industrial production" (l.c., p. xvi). However, they also point to the good news themselves: "The decline is not inevitable. To avoid it, two changes are necessary. The first is a comprehensive revision of policies and practices that perpetuate growth in material consumption and in population. The second is a rapid, drastic increase in the efficiency with which materials and energy are used." (*ibid.*)

They are right in saying that efficiency won't be enough. If exponential growth goes on at a rate of 5 % per annum, the entire factor four efficiency revolution would be eaten up within less than 30 years!

The Meadows and Randers give a fairly convincing, if depressing, answer to the question of why politics performs so poorly on preparing for the 21st century. In the following graph, the term of office for political leaders elected in 1990 is juxtaposed to the expected pay-back period for major corporate investments, the operating life of a power plant (the paragon of long-termism in the business world), the lifetime of a child born in 1990, the influence on the ozone layer of chlorofluorocarbons manufactured in 1990 and finally the time horizon of their World3 Model.

11.6.2 Population Dynamics

None of the problems discussed at the Earth Summit would be serious if we had only some 500 million people to feed, clothe and shelter. Oddly, however, the population issue was not even a topic for discussion at UNCED. Rumours went that this was a concession to the Vatican and to certain Islamic countries. It is hard to believe that religious ideas should prevent people and the countries they live in from doing what is needed to preserve God's creation and a decent livelihood for humans on this earth. But there we are. Some religious leaders seem to fear that population policies may lead to a general erosion of the respect for life.

World population is increasing at a rate nearing 100 million people per year. The developing countries account for 95 % of the increase. On the other hand, every additional US citizen statistically adds more stress to the natural environment than 20 Indians or Bangladeshis. From an ecological point of view, meaning in terms of "ecological footprints", most Northern countries are far more overpopulated than India or China. Correctly, the President's Council on Sustainable Development (PCSD 1996) put high emphasis on avoiding unwanted births in the USA.

A medium estimate by the United Nations projects total world population at 10 billion people by the year 2050. The high estimate is at 12 billion, and the low one at 8 billion.

Notwithstanding certain political and religious objections, the international community *did* address population questions. The United Nations convened the ICPD the International Conference on Population and Development, in Cairo in August 1994. At this Conference, the third of its kind, very valuable contributions were made towards rational population policies. In particular, the ICPD emphasised the importance of the role of women, their status in the society, their education level and their financial independence. The 1990 UNDP Human Development Report strikingly documented how population growth in 10 developing countries correlates with the absence of female education.

11.6.3 What has 'Factor Four' Got to Do with the Population?

Assuming per capita consumption increases of a mere 1.5 % per year (China maintains annual increases of some 8 % for years!), the medium variant would lead to a quadrupling of total consumption from 1995 to 2050. In other words, the total factor four revolution"—if it took place during that time span—would already be eaten up by this double dynamics of population increase and a very modest increase of consumption. Nothing of the efficiency revolution would be left for a relief to the over-stressed natural environment. How much worse would it be if the efficiency revolution were *not* taking place!

Politically speaking, the dramatic increase of population would almost unavoidably lead to conflicts over land and resources. Migration would reach all continents and countries. It is very much in the interest not only of the poor, but also of the rich, to halt population growth and eventually to make the low, 8 billion people scenario come true.

We ought not to introduce the factor four revolution into the population discussion in a merely defensive fashion. It could actually play a decisive role in reaching a stabilisation of world population. It has been a well established demographic fact for more than 50 years that populations tend to stabilise, regardless of religion, when a certain degree of prosperity is reached (which also correlates highly with the independence and self-esteem of women, although that latter correlation is also dependent on religion and culture).

If we now learn that American style prosperity for ecological reasons is *definitely not* conceivable for six or more billion people, then the hopes for stabilising population the 'natural' way are very dim. If, however, the efficiency revolution allows prosperity to occur at resource consumption levels roughly a factor of four below America's, we may become hopeful again.

In other words, those in the North who consider population growth the biggest menace, should do all they can to facilitate both prosperity developments and the efficiency revolution in the South. And as the South is not going to embark on the efficiency revolution on its own, it is all the more urgent for the North to begin the new trend!

After having briefly discussed the central factor of the Meadows' system dynamics, population, we may now turn our attention to the question of whether there are scenarios available that move beyond the purely educational value of the "start sustainability policies in 1975" scenario. Is there a reality-based scenario for a sustainable future?

Global scenario

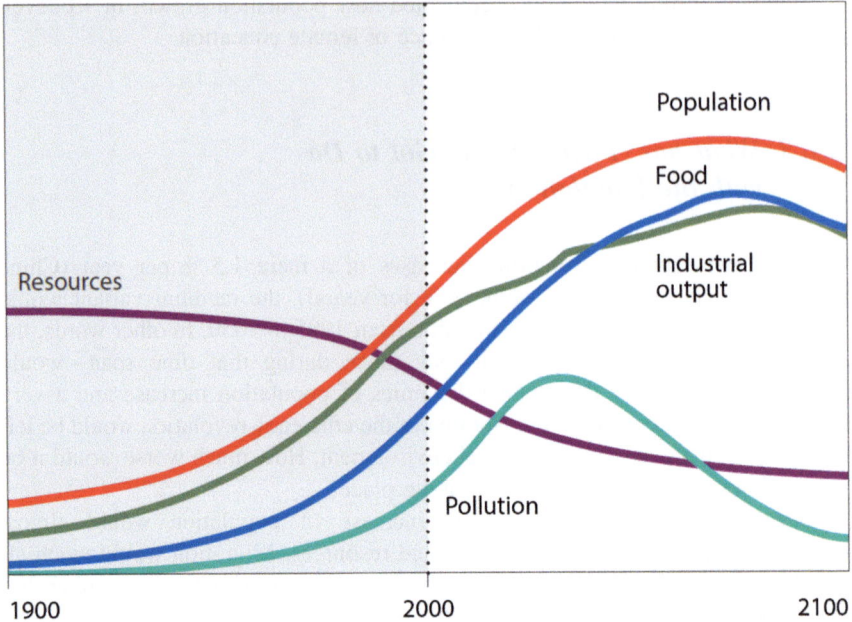

Fig. 11.3 Assuming 4 % annual gains in resource productivity, stabilization can be expected after 2050 on a higher level of prosperity. The authors gratefully acknowledge Dennis Meadows' willingness to share his software with us and Harry Lehmann's talent in injecting the factor four idea into the World3/91 model

We address this question by applying the World3/updated 1991 systems dynamics model, which Dennis Meadows has kindly made available to us.

The main idea is to inject the dynamics of the efficiency revolution into the existing model. We assume that efficiency gains of 4 % annually can be obtained. From the prosperity and education parameters we expect a decrease in family size, i.e. birth rates by 40 % from 2000 to 2100. As a result, a truly attractive scenario emerges from the 4 % gains assumption (Fig. 11.3). The 21st century need not be depressing at all. If our 'neocornucopian' visions come true, even the gravest world-wide distribution problems can be solved without any part of the world having to accept significant sacrifices in their well-being. What cannot be dealt with by globalised computer models are wars and other conflicts. Also, irrational behaviour under the permanent stress of world-wide economic competition cannot be accounted for.

References

Giarini, Orio; Stahel, Walter R., 1989/1993: *The Limits to Certainty—Facing Risks in the New Service Economy* (Boston, Mass.: Kluwer Academic).

Gore, Al, 1992: *Earth in the Balance. Ecology and the Human Spirit* (Boston: Houghton Mifflin).

Maddox, John, 1972: *The Doomsday Syndrome* (London: Macmillan).

PCSD (President's Council for Sustainable Development), 1996: *Sustainable America: A new consensus for prosperity, opportunity and a healthy environment in the future* (Washington, D.C.: PCSD).

Paul Hawken; Lovins, Amory B.; Hunter Lovins, L., 1998: *Natural Capitalism* (London: Earthscan Publications Ltd).

Schmidt-Bleek, Friedrich, 1994: *Wieviel Umwelt braucht der Mensch—Faktor 10* (Munich: DTV).

Chapter 12
Eco-Efficiency Goals: Factor Four or Factor Ten

12.1 Why Should Factor Four and Factor Ten Co-exist?

When the Wuppertal Institute (which was founded in 1991) was in its baby phase, I observed two mutually conflicting debates going on: (i) the preparations for the Earth Summit of Rio de Janeiro and (ii) public and business debates about *business as usual* environmental policies. The latter essentially circled around pollution control and certain efficiency improvements of some five or ten or maximum 20 %.[1]

Taking a closer look at the challenges emerging in the context of the Earth Summit we felt at the Institute that they had rather little to do with pollution control and that ultimately *dramatic* increases of resource productivity were going to be needed to respond to the challenges. This holds for all three major Rio themes:

(i) Biodiversity: At present, we are losing some twenty, perhaps fifty plant and animal species every day. This is mostly due to the destruction of natural habitats resulting from land conversion for mining, farming, housing, infrastructure etc. Perhaps the most important reason for habitat destruction is the gigantic flows of materials that are induced by our modern consumer society. Each one of us in the North induces material flows of some 50 tons/year (Schmidt-Bleek 1994). A major reduction in material flows is needed worldwide. Developing countries, on the other hand, want growth (and the North too, to be honest). If both goals, massively reducing material flows and economic growth, are to be reconciled, an absolutely breath-taking increase of resource productivity will be required. A factor of ten is a fair guess.

(ii) Climate change: Climatologists recommend a reduction of greenhouse gas emissions by some 50–80 % by the middle of the next century. Let us hope that the lower figure, 50 % will do. On the other hand, the demand for energy services is likely at least to double within that period. Increasing energy (or, rather, carbon) productivity by a factor of four would greatly help to close that gap.

---

[1] This text was first presented to an OECD Workshop in Paris, 3–4 September 1997 on Fostering Eco-Efficiency: the Role of Governments and has not previously been published.

(iii) Sustainable development: The Northern style prosperity generally rests on very high levels of resource consumption, easily ten times higher on a per capita basis than those of the less developed countries. In the language of William Rees/Matthis Wackernagel (1994), the 'ecological footprints' of US citizens, Germans or Japanese are some ten times larger than those of Chinese and Indians. In this footprint language, the USA, Germany and Japan are hopelessly overpopulated, while China and India are not. Increasing resource productivity by a factor of four or six or ten seems like minimum precondition for achieving sustainable development world-wide.

If my initial question was why should I allow 'factor four' and 'factor ten' to co-exist in one institute, the answer is: 'factor four' or 'factor ten' is the *right* question to be asked, while '+10 %' or '+20 %' increase of resource productivity is the *wrong* question.

12.2 Which Are the Differences Among the Two?

If I *want* the public to be puzzled about the co-existence of factor four and factor ten, I feel nevertheless obliged to explain the difference of the two goals. For that I shall give three different, although mutually supportive answers:

(i) Assuming, on a macro-economic level, an annual increase of 5 %, we will reach a factor of four in 28 years, and a factor of ten in some 45 years.
(ii) If you want to encourage (rather than discourage) the public, notably the engineering and business communities for your policy goals, you have to give lots of exciting and realistic *examples* of how the goal can be pursued and reached. Such examples have to be *sufficiently narrowly defined* to make sense in product design inside one company. I found it relatively easy to collect—together with Amory Lovins—50 examples of quadrupling resource productivity (Weizsäcker et al. 1997), while my attempts to find plenty of examples of decupling it largely failed. A factor of ten can often be reached if complex life-cycle considerations are used, but typically not the entire life cycle is in the hands of one company. Also, long term macro-economic considerations can rather conveniently lead to a factor of ten, but the line of proof is complex and is not exactly encouraging those who remain doubtful about the whole approach.
(iii) In particular, if you want to address energy productivity, in very many instances a factor of four is available while a factor of ten is not. Take the fuel efficiency of cars. From an average 8 L/100 km you can—with courageously stretched imagination—arrive at Amory Lovins' 'hyper car' with an assumed fuel efficiency just below 2 L/100 km; but 0.8 l/100 km is simply outlandish. The same holds for air conditioners, light bulbs and refrigerators. You are often running against the Second Law of thermodynamics if you want more. In material efficiency, on the other hand, extending product lifecycles and recycling metals hardly meets limitations lying in the laws of physics.

12.3 Which Are the Appropriate Time Horizons?

By asking the question of time horizons I try to do away with one public misconception from the beginning: Factor four or factor ten is not something governments can (or should be allowed to) prescribe within the typical reach of their legislative time horizons. When we talk about factor four or factor ten, we are talking about one or two generations, i.e. 30–60 years.

My simplistic advice in terms of practical policy is: Avoid capital destruction and make use of the 'natural' rate of exchange of the capital stock. But try to make sure that fresh capital is no longer invested in dinosaurs but in the new generation of efficient and superefficient technologies.

One the other hand, there may be non-linear events of a dramatic deterioration of the environmental condition, such as the gliding into the Sea of considerable parts of the Antarctic or Greenland ice shelves. If scientific fears are building up for such events to become imminent, evidently the advice outlined above about no enhanced destruction of the capital stock will collapse.

12.4 What Should be Done Next?

My first plea is to all actors to recognize that thinking in terms of 'factors' is distinct from classical environmental policy. In terms of government portfolios it rather relates to the economics, technology and finance departments than to the environment department. (Which doesn't make things easier.)

For the business community it is important to recognize that considerable opportunities already exist to increase profitability by increasing resource productivity (Blumberg et al. 1997). Resource productivity should in any case receive higher attention and emphasis in the context of ecological auditing (EMAS and ISO 14000).

Special events such as the holding of a Factor Four + Fair in Klagenfurt, Austria, June 1998, the inauguration in the USA of the *Future 500* or the founding in both Cantons of Basle of a Factor Four Club can help increase public awareness for what is now on the agenda. In particular, the founding in 1994 of a Factor Ten Club by Friedrich Schmidt-Bleek and his recent creation of a Factor Ten Institute at Carnoules, France, have met with major international interest.

As I see it, the most important thing to do in the context of the 'Factor agenda' is to realize that we are heading for a new direction of technological progress. In the past, the main emphasis in technological progress was laid on the increase of *labour productivity* which may have risen 20-fold during the last 150 years. This emphasis was very reasonable 150 years ago when human labour was very inefficient indeed, and very hard too. Nature, on the other hand, seemed available nearly unlimited. So the exploitation of nature seemed like a legitimate and natural part of the game.

But today it's a different situation altogether. Labour is abundant, labour productivity is high, and the real scarce resource is nature. This means it is now high time to concentrate our efforts on the increase of resource productivity. Slowing down the increase of labour productivity while speeding up resource productivity should make countries richer, not poorer which have high levels of unemployment and have to import much of the natural resources needed; that's the typical situation of most OECD countries.

Redirecting technological progress is, of course, a 'century task' and will involve all kinds of actors, engineers and managers, auditors and analysts, government officials and members of parliament, educators, journalists and NGO activists.

12.5 Which Can be the Role of Governments?

This Conference is chiefly addressing the role of governments. What can governments do to make the transition happen with maximum benefit and minimum losses? I should like to repeat that governments should not prescribe any 'factor'.

Rather, governments should work on the *frame conditions* in the sense that resource efficiency becomes more profitable. World market prices typically tell the world the untrue story of an ever fuller cornucopia of resources. Intensified exploration and technological progress have made access ever cheaper even when exhaustion comes nearer. Hence some upward price correction is justified to reflect long term scarcity and environmental costs. Ironically, however, government action is typically working in the opposite direction. Massive subsidies amounting to some 700 billion dollars are given to encourage resource use (De Moor et al. 1997). Here, clearly, it will be urgent for governments to stop and reverse such perverse incentives.

Reversing the perverse incentives chiefly means the (gradual) introduction of an ecological tax reform. The fiscal burdens on human labour should be reduced, those on the use of natural resources increased. But to honour the above principle of avoiding capital destruction, compromises have to be designed to protect the capital stock even of very resource intensive industries. Denmark and The Netherlands have shown the way, I believe.

Other government actions relate to public procurement. The purchasing power of the state can make a huge difference on product and service markets.

For governments, there is also scope for regulatory action. Good examples are the Swedish building code, which leads to a cutting by roughly a factor of three of the energy needed for heating, and the new German recycling law, which encourages durability, reuse and recycling, i.e. material productivity. However, if we want to address the secular changes indicated above, the redirection of technological progress, nearly all government departments will be involved. In particular, all incentive structures working on the profitability of certain technological investments will have to be reconsidered. A new R&D strategy will be needed including for universities and technical colleges. And industrial norms, regulations and unwritten customs will have to be scrutinized under the challenge of the technological transformation.

If one country goes ahead and is successful in the new arena it may have an enormous influence on world-wide competitiveness and thereby exert a considerable pull for other countries to follow.

References

Blumberg, Jerald; Blum, Georges; Korsvold, Åge, 1997: *Environmental Performance and Shareholder Value* (Geneva: World Business Council for Sustainable Development).
De Moor, André; Calamai, Peter, 1997: *Subsidising Unsustainable Development; Under-mining the Earth With Public Funds* (Toronto: Earth Council).
Rees, William; Wackernagel, Matthis, 1994: "Ecological Footprints and Appropriated Carrying Capacity: Measuring the Natural Capital Requirements of the Human Economy", in: Jannson, A.M.; et al. (Eds.): *Investing in Natural Capital: The Ecological Economics Approach to Sustainability* (Washington: Island Press).
Schmidt-Bleek, Friedrich, 1994: *Wieviel Umwelt braucht der Mensch?* (Basel: Birkhäuser).
Weizsäcker, Ernst von; Lovins, Amory; Lovins, Hunter, 1997: *Factor Four. Doubling Wealth, Halving Resource Use* (London: Earthscan).

Chapter 13
Sharing the Planet: From 'Limits to Growth' to 'Factor Four'

13.1 From 'Limits to Growth' to Sustainable Development

In the organic world nothing is more commonplace than limits to growth. Even the tallest redwood pines stop growing at some point.[1] And even the most aggressive bacteria stop proliferating—when they reach the limits of their host range. As a matter of fact, it is 'wise' for parasites to let their hosts flourish,—which also implies a degree of restraint.

Only humans, during a few centuries of 'discovery' and conquest adopted the habits of never-ending growth. When economics assumed the role of a world religion, which happened after 1945, the idea became canonized that there are simply no limits to growth.

All the more shocking was the publication in 1972 of the Club of Rome's *Limits to Growth*. Shocking it was, but it was impossible for the public to ignore or discard it. The Club was simply too prestigious and the Report to suggestive through its use of state of the art computer modelling. The Report written under the supervision of Prof. Jay Forrester by Dennis Meadows, Donella Meadows, Jørgen Randers and William W. Behrens became a milestone indeed in the international debate on the future of mankind.

The Report said that catastrophic collapses of the Earth's life-support system would occur during the first half of the 21st century if growth trends were continuing unmitigated. Alas, the report was published at a time when, after the big wave of decolonialization, the developing countries forcefully appeared on the political arena and understandably found it scandalous that some intellectuals from the North were precisely now proclaiming the 'limits of growth'. To them, this was like an extension of colonialism.

---

[1] This text was first published in a text for the Dutch association of the *Club of Rome* edited by Bob van der Zwaan and Arthur Petersen on: *Sharing the Planet; Population-Consumption-Species; Sustainability of the Planet and International Politics: Looking for an Effective Global Governance* (Delft: Eburon Publishers, 2003). The permission to republish this text was granted by Prof. Eric-Jan Tuininga on 24 February 2014.

One major report was written to confront 'Limits to Growth', the 'The Bariloche Model' by Amilcar Herrera and others that was pinpointing the alarming gap between rich and poor countries. Other critics related to methodological weakness of the Report and its 'World 3 scenarios':

(i) Resource depletion. Economists say that the 'reach' of depletable natural resources has always and for systematic reasons been in the vicinity of 30 years, because this is about the time span worth looking at for companies and states concerned with resource availability. They always look at the low hanging fruits and simply don't go into the trouble and costs of developing access to the higher hanging ones. Hence, economists considered the Meadows assumption of a 30 years reach of gas, oil or a copper a trivial statement on which no depletion forecasts should be built.

(ii) Pollution. In 1972, pollution was the most visible environmental threat. A constant mathematical relation was used in the World 3 computer model between pollution and industrial output. This relation, obviously, was wrong. For the simple reason that Western societies were able to answer the challenge with and pollution control technologies. Pollution control at the end of the pipe was actually very convenient for the business camp. Companies could always argue that only if they were prospering they could do the anti-pollution job properly. The whole game of pollution control ended up in the 'inverted U curve' paradigm: Societies start poor and clean. In the process of industrialization they become rich and dirty. When they are rich enough to combat pollution, they finally become rich and clean. That was a lovely and harmonious paradigm, squarely contradicting the gloomy picture painted by the Club of Rome!

(iii) The static relations between the five main factors used in the World 3 scenarios of the Limits to Growth Report. What was true for pollution control almost equally holds for the other parameters used by the Meadows team, namely population, food, resources, and industrial output. Technological progress can be said to consist in the de-coupling of such parameters. As prosperity grew, the dynamics of population growth came to a halt, as can be seen happening dramatically in countries like Japan, Italy, Spain and certain Latin American countries, not to speak of the North European countries where this has been known in 1972 already. Also the food-industry-nexus has always been subject to change, so far only in the favourable sense.

If that is so, why am I still defending the Limits to Growth Report? The reason is plain and simple. Because the main thrust of that pivotal Report is not at all touched by the adjustments which can take care of the three objections mentioned. This world is a limited one. Human societies must not and cannot grow beyond the natural limits which ultimately do apply to the resource base as well as to the food base. However the nature of the ultimate limits can be seen and described differently from the World 3 scenarios.

The new language, introduced by the Brundtland Report of 1987, refers to *sustainable development*. At the Earth Summit held 1992 in Rio de Janeiro, the Agenda 21 was adopted to outline the immense tasks of sustainable development. Also two major environmental conventions were adopted, the Climate Framework Convention and the Biodiversity Convention.

In terms of Agenda 21 it can be deduced that the harmonious 'inverted U curve paradigm' in itself is *not* sustainable. That is because 'rich and clean' in its present meaning involves *per capita* consumption levels of depletable resources easily 5–20 times the rate of the 'poor and clean' stage. These depletable resources include also the absorptive capacity of the atmosphere. And the stage of depletion will be reached rather rapidly if 6 or 10 billion people become 'rich and clean'.

Relating to the sustainability challenge, Matthis Wackernagel and William Rees (1997) have introduced the concept of the 'ecological footprints' and they show that US citizens, Germans or Japanese have some ten times larger footprints than the Chinese or the Indians. In this footprint language, the USA, Germany and Japan are hopelessly overpopulated, while China and India are not.

Not yet. The developing countries, assisted by international investors, work very hard to *develop*, i.e. to emulate Western styles of industrialization and of consumption and thus to acquire Western size footprints. By 2030, also China and India will be hopelessly overpopulated in terms of their footprints. We would need three to four planets Earth to accommodate 6–8 billion US size ecological footprints. That is a rather drastic way of demonstrating that our present Western lifestyles are ecologically *unsustainable* and collide with the limits to growth.

13.2 Biodiversity and Dematerialization

Let us now take a look at some of the characteristics of the real limits to growth and of the Western size footprints. If it is not air or water pollution and if it is not any imminent depletion of gas or oil or copper, what is it then that causes so much is the ecological harm that we should speak of real limits?

Among the most alarming effects of civilization and economic growth is the rapid *loss of biodiversity*. At present, we are losing some 20, perhaps 50 plant and animal species every day. This is mostly due to the destruction of natural habitats which have been the home to hundreds of thousands of biological species, some of them rather inconspicuous but nevertheless important in the interlinking webs of ecosystems. Habitat destruction mostly results from land conversion for mining, agricultural use, forest monocultures, or settlements. Developing countries tend to export most of the products of their lands. This is how we in the industrialized countries are able to maintain total footprints exceeding our own territories. We 'export' much of our footprints to the South.

One reason for the massive land conversion and habitat destruction, perhaps the most important reason, is the gigantic *flows of materials* which are induced by our modern consumer society. Each one of us in the North induces material flows (or 'ecological rucksacks') of some 40 up to 80 tons/year.

I cannot see any plausible strategy of protecting the remaining biodiversity without *drastically* reducing the material flows travelling through the human technosphere. However, the Biodiversity Convention adopted at Rio de Janeiro does not even mention the nexus to dematerialization.

How much dematerialization do we need if we want to allow developing countries to reach our levels of prosperity while simultaneously reducing the pressures on land for wild life and biodiversity? Rough estimates by my friend Friedrich Schmidt-Bleek (1997) suggest that we shall need at least a factor of ten in reducing the ecological rucksacks in the West. That's a huge challenge both to technology and to our civilization.

13.3 Climate Change and the Energy Dilemma

Materials is one part of the story. The other is energy. A sizeable part of the 'footprints' are actually not left at the ground but blasted into the air: it is human-caused greenhouse gas emissions. We are significantly changing the chemical composition of the atmosphere. By 2030, the carbon dioxide concentrations will have doubled as compared to pre-industrial levels. Insurance companies, notably reinsurers fear further increasing storms and floods. Annual damages have already exceeded US$ 50 billion. If climate develops further as some climatologists foresee, the countries worst hit will be developing countries, not least the small island states which in the worst case can be literally washed away.

The scientific basis for such fears lies in the famous correlation between CO_2 concentrations and temperatures, discovered by excavations from the Antarctic ice of air bubbles up to 160,000 years old. More alarming is the correlation between temperatures and a *third* parameter, the sea water table which can vary by some 200 m. The geography of the coast lines, therefore, has changed dramatically in geological times. Theoretically, the flood can come in a matter of a few decades. According to Michael Tooley (1989), some 7800 years ago the better part of the ice masses over Labrador and the Hudson Bay were breaking off into the sea letting the global sea water table rise by some 7–8 m. I am not suggesting that anything of this kind is *likely* to happen during the next 50 years. But we have no certainty that it will not happen.

What can we do to stop the dangerous trends of climatic change? Climatologists recommend a reduction of greenhouse gas emissions by some 50–80 % by the middle of our century. This would enable us to stabilize CO_2 concentrations at present levels. On the other hand, we learn from the World Energy Council, that the demand for energy services and with it the emissions of carbon dioxide is going to rise steeply and is most likely at least to double within that period. So that's at least a gap as large as a factor of four which will have to be closed.

Some people say we can close the gap by turning to nuclear from fossils. But today, nuclear is a mere 7 % of the world energy pie. Even this is subject to severe conflicts, and only small parts of the risks are covered by private insurance contracts. Imagine a neck-breaking rush towards tripling nuclear energy supplies in 40 years,—a political nightmare given the vulnerability of all installations to terrorism and war. What we would gain is an increase from 7 to 21 % of the pie. But while the pie itself is doubling, we are falling back to a mere 10 %. That's doesn't seem to be the master key solution to the climate challenge.

With renewables, the substitution of fossils is a lot nicer but almost equally frustrating. Wind and solar make up less than 1 % of the present pie. Let's assume an heroic strategy of increasing it 20-fold. Then we have reached 20 % of the present pie, but a mere 10 % of the double sized pie. Hydro has more at present, but please remember what nightmares are associates with present-day hydro schemes such as the Three Gorges Dam in China. We conclude this section by plainly stating that energy policy too is in a massive dilemma.

13.4 After the Industrial Revolution the Eco-Efficiency Revolution

The challenges of sustainability, of biodiversity protection and of climatic change look breathtaking. Lifestyle styles with unsustainably large footprints and yawning gaps of factors between four and ten in the fields of energy/climate and of materials flows/biodiversity could leave us rather helpless. Fortunately, there is hope. Much of this hope is rooted in technological progress. But the task will be no smaller than the adventure of the Industrial Revolution.

What kind of animal is technological progress? We seem to assume, all of us, that technological progress is some undirected 'natural' phenomenon that comes out of mix of scientific ingenuity and economic competition. States are said to have at best a chance to accelerate it or to impede it by bureaucracy or by setting unenlightened priorities. This standard picture of technological progress, I believe, is profoundly wrong. Technological progress has got a direction that can be understood and steered.

In the past, technology was mostly driven (if not by military considerations) by the desire of economic expansion. The main emphasis was laid on the increase of *labour productivity* which may have risen 20-fold during the last 150 years. That increased labour productivity becomes visible in the speed of our vehicles, in the power of our machines, in the organizational miracles of industrial production lines and in the unprecedented skills of modern information technologies.

The emphasis on labour productivity was very reasonable 150 years ago when human labour was very inefficient indeed, and very hard too. The winners in economic competition were almost always those who could offer more services and goods with less human labour. Wages rose more or less in proportion to the increases of labour productivity. So workers were well advised to support ever further productivity increases.

Nature seemed to be available nearly unlimited. So the exploitation of nature seemed like a legitimate and natural part of the game. The game was later called by historians the Industrial Revolution. And it is still going on world-wide.

Today, however, we are living in a completely different world from the early 19th century. Today labour is abundant, labour productivity is very high, and the real scarce resource is nature.

This means it is now high time to concentrate our efforts on the increase of *resource productivity*. Even purely economic—and social—reasons speak for it. Slowing down the increase of labour productivity while speeding up resource productivity should make countries richer, not poorer which have high levels of unemployment and have to import much of the natural resources needed.

Unemployment is a worldwide phenomenon. According to figures of the *International Labour Organisation* (ILO), there are roughly 800 million people jobless. This is a tragedy for their families and a disaster for public budgets in those countries in which the state is obliged to pay unemployment benefits.

Shifting emphasis to resource productivity should be the best answer also to the aforementioned challenge of sustainable development. The World Business Council for Sustainable Development is usually speaking of Eco-Efficiency as new guiding term. According to the considerations made above, eco-efficiency should go up by at least a factor of four. Thus we aim at productivity jumps equally impressive as there were characteristic of the Industrial Revolution. So let us therefore speak of the Eco-Efficiency *Revolution*.

The Eco-Efficiency Revolution is perhaps the only strategy allowing a reduction in size of the ecological footprints without jeopardising employment and competitiveness.

13.5 The Good News: Factor Four

Before entering the competitiveness questions let us have a look at what is possible today in terms of dematerialization and energy productivity. Here I have some good news. *It is possible at least to quadruple resource productivity*. Our 1995 Report to the Club of Rome, *Factor Four,* features 50 examples for increasing resource productivity by a factor of four at least. Twenty examples were selected in the field of energy, 20 in material resource productivity, and ten in transportation.

The German version of Factor Four was published in September 1995 and became a best-seller immediately. Twelve translations exist including Russian, Chinese and Japanese editions. After the Limits to Growth, *Factor Four* has probably been the report to the Club of Rome reaching the largest number of readers world-wide.

13.5 The Good News: Factor Four

Let us have a look at the substance of the Report. One very attractive example is what co-author Amory Lovins has dubbed the *hypercar*. By almost entirely redesigning cars, making them light-weight and still crash-resistant, and by using modern hybrid engines, the average fuel consumption can be pushed below 2 L/100 km, which is more than four times better than today's fleets.

A few examples relate to the energy use of both private homes and office buildings. High tech insulation both of walls and of windows and an efficient heat exchange ventilation can reduce heating requirements by as much as a factor of ten.

Other examples include light bulbs, refrigerators, air conditioners, TV sets, mechanical fans, pumps and motors, computers and other office equipment. One is the success story of energy and waste savings over 10 years at a big chemical firm in the USA with astounding average returns on investment above 100 % per year.

Also renewable sources of energy will play an important role in the efficiency revolution. They may not by themselves save energy. But they are at least "carbon-efficient", and they lend themselves to being combined with efficiency technologies, e.g. in the case of passive solar energy in buildings which is optimized in the so-called translucent insulation technique.

A different and very important sector of energy use is nutrition. By reducing the excessive use of fertilizers and the transportation of fodder, and by slightly cutting meat consumption, energy requirements for a healthy diet can be cut by a factor of four.

The 20 examples of revolutionising material productivity range from construction and durable office furniture to water in homes and in paper manufacturing and to high tech recyclable plastics for wrapping and catering.

Both in energy and in materials efficiency one striking feature is the strengthening of the crafts sector in our economies. Insulation of homes, renewable energy, manufacturing and repair of durable goods, they all are crafts-based. Needless to say, modern craftsmen are well equipped with modern communication technology. One example is tele-repair services allowing the expert via a TV—connection to instruct customers in real time to repair a broken dish-washer or other appliances.

One fine example is the replacement of a clumsy paper-based filing cabinet by a modern CD ROM system. There you save more than a factor of ten even if you generously include the "ecological rucksacks" of the metals contained in the disks.

Similarly, the transport both of people and of goods can to a certain extent be replaced by information flows. Video conferences can—at least theoretically—save a lot of business travel. And e-mail needs much less resources than what is meanwhile called the "snail mail".

On the other hand, there are major rebound effects to be expected. If you first 'meet' your business partners on the screen and your contacts are successful you are more likely than before wanting to see them in person. Thus each video conference can be the cause of *additional* overseas travel.

Other examples from the transport sector relate to high tech measures increasing the capacity of existing railways and the cutting of ton kilometres for the production of yoghurt or fruit juices.

We did a calculation about the effects of efficiency on the World 3 scenario. Leaving the mathematical relations as the Meadows team has stated them but injecting 2 and 4 %, respectively, of annual efficiency gains we found an harmonious stabilization by 2100 or even by 2050, respectively. Thus, Factor Four can be considered a true answer to the challenges set out in Limits to Growth.

13.6 Profitability, Long Term and Short Term

Needless to say that much of the efficiency revolution is not going to happen unless the frame conditions for doing business will be changed. Efficiency must be made profitable.

However, to an astonishing extent, eco-efficiency is profitable now. Companies undergoing the eco-audit procedures or even without them paying sufficient attention to the resource flows going through the firm, have discovered that they gain considerable transparency also on the financial flows; they enjoy better cohesion with their staff; and they experience better customer relations. All this has led to the astonishing and most promising experience that portfolios of 'green' stocks can perform better on the stock markets even than the Morgan Stanley Capital International index which is the benchmark index for shareholder value.

It is to be feared, however, that the potential for making profits by eco-efficiency measures will be narrowly limited if the present world market conditions prevail. These are characterized to a large degree by the widespread obsession with classical industrialization and with the idea of local politicians that each and every industrial investment deserves their special attention and support. As a result, we see the most incredible amount of subsidies going into resource eating activities. As André de Moor (van Beers; de Moor 2002) has estimated, some 700 billion dollars are spent annually in the four fields of energy consumption, water, agriculture and motor transport. This does not yet account for all the tax advantages, free infrastructure and land given to the investor. De-subsidising resource use will be an important policy world-wide. But like in the case of pollution control, one country can hardly move if the competitors don't.

Another, and related, policy tool is ecological tax reform. In a world of growing unemployment and of scarce natural resources it just doesn't make sense to draw the biggest part of fiscal revenues from human labour while resource use goes essentially free of charges. Ecological tax reforms are applied by the majority of the EU countries.

Such new incentive systems directed towards long term profitability and sustainability, should be honoured by higher profits once our Western societies (and in their wake others) begin to realise in a broad and pervasive manner that the present technological trends are unsustainable. Realising that a new technological revolution is waiting around the corner, enlightened states and business communities can initiate a rat race in that new direction. Then it will be the eco-efficiency pioneers harvesting the first mover advantages.

So I may end with a rather optimistic outlook. The de-coupling of well-being from resource use can go on rapidly. It will be both private and public actors who can play important roles in accelerating the transition. And the Club of Rome is extremely well positioned between the sciences, the states and the business community worldwide to create a powerful and lasting momentum driving our spaceship in the new and sustainable direction.

References

Blumberg, Jerald; Blum, Georges; Korsvold, Åge, 1997: *Environmental Performance and Shareholder Value* (Geneva: World Business Council for Sustainable Development).
Fussler, Claude; James, P., 1997: *Driving Eco-Innovation* (London: Pitman).
Meadows, Dennis; Meadows, Donella; Randers, Jørgen; Behrens, William, 1972: *The Limits to Growth* (New York: Universe Books).
Meadows, Dennis; Meadows, Donella; Randers, Jørgen, 1992: *Beyond the Limits* (Post Mills, VT.: Chelsea Green).
Schmidheiny, Stephan; Business Council for Sustainable Development, 1992: *Changing Course*. (Cambridge, MA.: MIT Press).
Schmidt-Bleek, Friedrich, 1997: *Wieviel Umwelt braucht der Mensch?* (München: DTV).
Tooley, Michael J., 1989: "Global Sea Levels: Floodwaters Mark Sudden Rise", in: *Nature*, 342: 20–21.
Van Beers, Cees; De Moor, André, 2002: *Public Subsidies and Policy Failures: How Subsidies Distort the Natural Environment, Equity and Trade, and How to Reform them* (Cheltenham: Edward Elgar).
Wackernagel, Matthis; Rees, William, 1995: *Our ecological footprint* (Gabriola Island, BC, Canada.: New Society Publishers).
Weizsäcker, Ernst von; Lovins, Amory; Lovins, Hunter, 1997: *Factor Four. Doubling Wealth, Halving Resource Use* (London: Earthscan).

Ernst Ulrich von Weizsäcker, a former chairman of the Federation of German Scientists (Verereinigung Detuscher Wissenschaftler) discussion with Rainer Langhans, renowned commune founder of the late 1960s visiting the photo exhibition during the conference organized by the VDW on the 100th birthday of is father, Carl Friedrich von Weizsäcker. *Source* Photo was provided by VDW which granted permission to use it here

Chapter 14
Globalization and Its Challenges to Democracy and Development

Mr. Leonel Fernández, Ladies and Gentlemen,

I feel greatly honoured being invited by your foundation to give a lecture on globalization and its challenges to democracy and development. Originally, the idea was that I speak about "Factor Four", an exciting universe of technologies in the making, which allow a quadrupled resource productivity. This means that the world could double its wealth and simultaneously cutting its use of energy and material resources in half. But, honouring the central mission of your Foundation, I agreed to address the globalisation challenges.[1]

You may assume that globalisation has been around in this world since days of Cristobal Colón who landed not too far from here for the first time thereby 'discovering' what later was named America (Fig. 14.1).

In my role as Chairman of the parliamentary Select Committee on Economic Globalization I was surprised to learn that the term *globalization* is actually brand new. It would celebrate its tenths birthday this year,—if words can celebrate their birthdays at all. It began to play a role in public in 1993 as can be seen on the first picture:

The strongest reason for the sudden appearance of the term globalization has been the end of the Cold War. We all, with the exception perhaps of your Cuban neighbours, were very happy at the collapse of the Soviet empire and what happened afterwards:

- We were freed from the spectre of a Third World War;
- Democracy, free speech and free press spread throughout the world;
- Well-positioned and well-governed countries enjoyed exciting new opportunities, e.g. the US (owning the largest amounts of capital which could be placed at the places of highest profitability world-wide) and China (with its immense labour force, emerging high technologies and high discipline). Also the

---

[1] This talk was presented on 4 March, 2003 to the Fundación Global, Democracia y Desarollo, Santo Domingo, Dominican Republic and has not previously been published. Then opposition leader Leonel Fernández Reyna became President of the Dominican Republic again from 2004 to 2012.

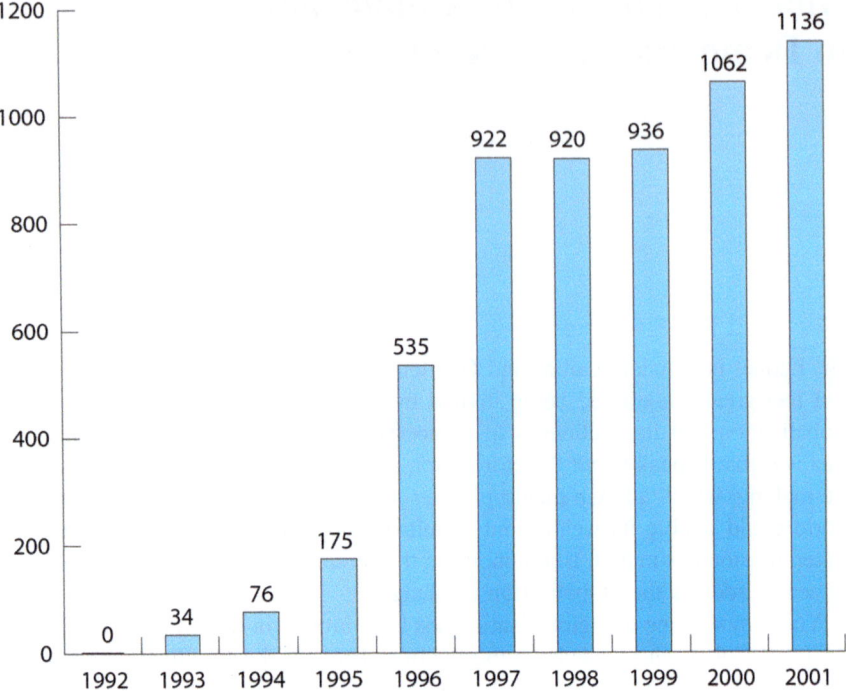

Fig. 14.1 The term *globalization* appeared in the languages of the world after the end of the Cold War. The *picture* shows the phenomenon for Germany's Frankfurter Allgemeine Zeitung. *Figure* Wuppertal Institute

Dominican Republic benefited at least since 1996, with Dr. Fernández assuming the presidency.

- Stock markets soared, letting market capitalization of the world's total stocks triple within 10 years.
- Inflation was sent down in most countries to the lowest levels since the 1950s.
- The Internet became an immensely powerful tool of worldwide communication.

However, for all the good news, there is also a *downside to globalization*. During the Cold War, international capital had always to seek consensus with national governments and parliaments in the North and the South. In the South, governments used to play on the East–West tensions to induce the inflow of official development aid or of private capital.

In the North, the Cold War forced governments to establish an attractive social security net to prove to the masses that capitalism took better care even for the poor than communism. In Germany it was the Social Market Economy launched by Ludwig Erhard in the 1950s. But the entire European Union was constructed around that model of an 'inclusive' capitalism. Progressive income taxes and hefty corporate taxation were never seriously disputed. For business and for the rich this may have been annoying but is was anyway better than any move of the country towards communism.

With the end of the Cold War, the need for consensus disappeared. Now the name of the game was catch as catch can. Competition got ever more brutal. You can see it from the number of business bankruptcies in many countries, including Germany.

Let me illustrate what happened by an anecdote from the automobile sector. Volkswagen got into deep troubles in 1993 because its cars were simply too expensive for the competitive world market. In response, they hired a fairly controversial Spaniard, Mr. Ignacio Lopez whose task it was to reduce costs which he did by squeezing the last penny out of the parts suppliers, by saying they had to supply the same quantity and quality of parts next year, but at 10 % lower costs. If they complained hinting at their own costs, he coldly replied that Volkswagen would then go to another supply firm, which may be in Malaysia or Czechia.

It sounds brutal and it was brutal. But then, Fiat, which, I am sure, was much gentler to its part suppliers, ran into deadly difficulties a couple of years later.

For the state, the situation was not at all more comfortable. The increasing weakness of democracy in negotiating with the private sector was soon felt in the field of taxation. Figure 14.2 shows the steady decrease of corporate tax rates, resulting from ever-increasing pressures the private sector imposed on the states, in this case the OECD states.

You can imagine that many people were truly happy with the new situation. The market philosophy originating with Adam Smith in the 18th century became something like a new religion. The Adam Smith Institute had a Christmas card recently showing the exuberant joy on the part of the new religion:

Unfortunately, there were also many losers. The world saw a growing gap between rich and poor. Since the 1970s, the factor of the accumulated income of the richest 20 % of the world population divided by the accumulated incomes of the poorest 20 % rose from 30 to 75!

During the 1980s, the reason for this growing gap were the 'forerunners' of globalization, the debt crisis and the "Washington Consensus", which encouraged the IMF to force liberalization, privatization and budget austerity on countries needing IMF money. In the mean time, Joseph Stiglitz and others have brilliantly demasked the Washington Consensus as something that mostly benefits capital and rather impoverished the poor. In Latin America the effects were aggravated by the skyrocketing since 1979 of the US dollar interest rates leading in heavily indebted countries to what is now called the *lost decade*.

Coming now to solutions let me first recapitulate that Adam Smith himself made it clear that at least three conditions needed to be fulfilled before markets could become a blessing for all. (i) External peace and (ii) internally the rule of law are necessary to let the "invisible hand" work for the wealth of nations. Moreover, (iii) the state has to safeguard services and investments, which are not by themselves profitable on the markets. Among them you would count mass education, infrastructure, social cohesion, culture or care for the environment.

Percentage of average corporate tax rates

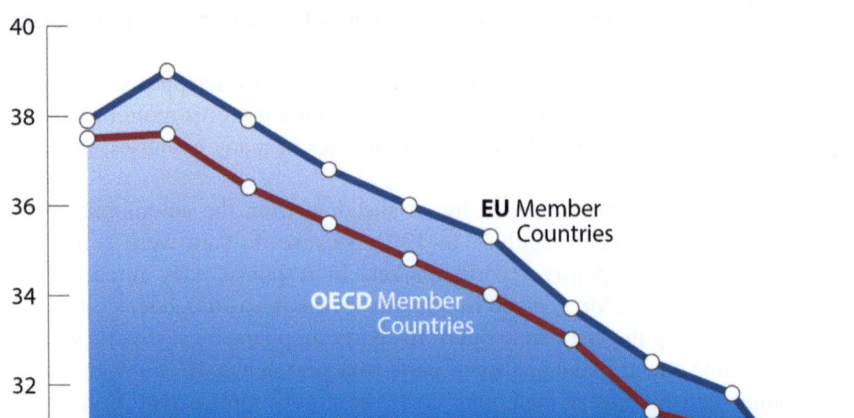

Fig. 14.2 Systematic decline of corporate taxation in OECD member states. The phenomenon results from tax competition among countries, one of the striking features of globalization. *Source* KPMG based on OECD figures

'Inclusive' capitalism can be seen as one of the best ways of fulfilling Adam Smith's conditions. When globalization began to attack social justice and the provision of elementary infrastructure services, it thereby began to destroy the basis of healthy capitalism.

Let me submit that it is now high time to re-balance public and private interests. In particular, the next trade negotiation in Cancún (WTO) and in Miami (FTAA or ALCA) may be pivotal at least for not surrendering more public interests to private free trade.

Let me illustrate my concerns with a well-known story that occurred in Canada under the NAFTA regime. A US-American chemical firm, *Ethyl*, began marketing a gasoline additive called MMT containing heavy metals. Researchers found hints that the stuff could cause cancer. As a result, Canada banned the chemical. Ethyl in turn sued Canada for destroying its expected sales, which was seen as a breach of the investment protection clause of NAFTA. Ethyl was likely to win according to NAFTA practice and Canada in order to avoid additional damage re-licensed MMT and paid Ethyl a compensation for estimated losses worth 100 million Canadian dollars.

What world are we in, where intergovernmental agreements protect private profits, not public health? May I suggest to the government of the Dominican Republic to apply extreme caution when the rules for FTAA are further negotiated in December in Miami? Investment protection may be a healthy thing to attract foreign direct investments, but central public interests must not be sacrificed in that context.

Investment protection is one of the "Singapore Issues" that the North will again put on the table in Cancún. Let me also advise your country to be extremely cautious on the Singapore Issues, which also include steps towards liberalization of public procurement and further competition rules, unless and until the North makes decisive and credible steps of opening their own markets for agriculture and textiles.

International trade regimes have been the most powerful instrument of cracking state authorities intending to protect their citizens, notably the poor ones. The General Agreement on Trade in Services, GATS, and the international patent agreement, TRIPS, both adopted in 1994 at the ceremony in Marrakech to establish the WTO and to end the Uruguay Round of the GATT, were seen by most developing countries as massive moves of strengthening the strong against the weak.

What happens if public water works are privatized and water becomes an expensive commodity for the poor? What happens if international seeds and drugs companies through TRIPS own the property rights of modern seeds and essential drugs?

I am not saying that trade in services and patent protection are undesirable, but we need a context of rules ensuring a reliable balance between pubic and private interests. How can that look like? Is there a chance of re-strengthening the nation state? I believe that is neither possible nor truly desirable (knowing unpleasant stories about state-cronyism and inefficiencies). What then?

Two mutually supportive strategies are available:

- International rule-setting, often called "global governance" but also rules-based regional group building such as in the EU:
- The strengthening of a *third actor*, *civil society*, which can case by case line up on public concerns with the state thus strengthening both.

The EU provides a reasonably attractive model. It has created over the years an ever growing intensity of economic cooperation and integration. But it has *simultaneously* created mechanisms of democratic control, legal supervision, geographical spread of the benefits accruing, and political coordination. We have a European Parliament, a European Court of Justice, the "Cohesion Funds" and a regular coordination of policies at the Council level. Soon we shall even have a European Constitution. All this, I am sad to say, is so far just unconceivable in the NAFTA context, chiefly because there is one NAFTA member that sees it as its privilege not to let anybody have a ruling on what it insists are its domestic affairs.

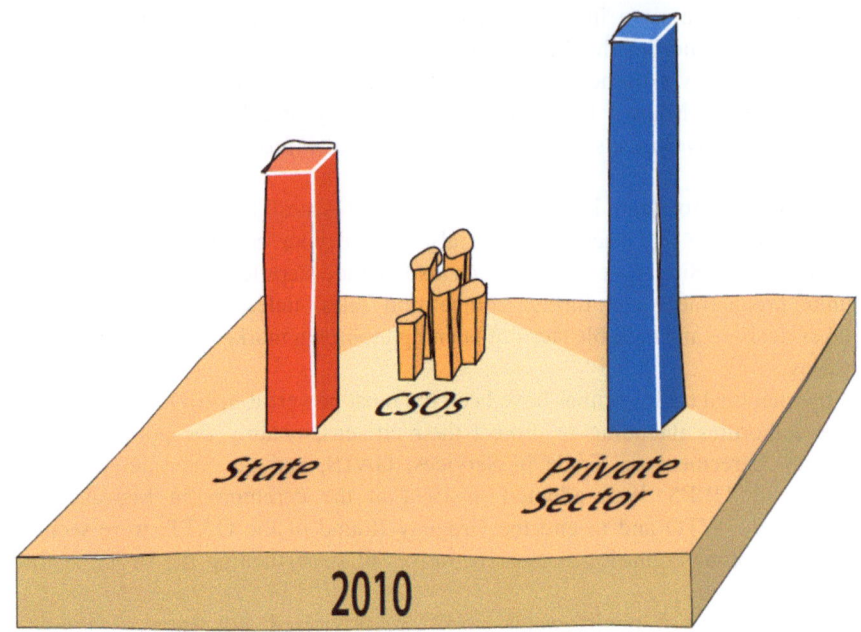

Fig. 14.3 Before 1990, powers rested mostly with the state. With the advent of globalization, the private sector began dominating. It is now high time to rebalance public with private goods. The Civil Society, as a third power structure, can greatly help doing that (*picture* Wuppertal Institute)

Clearly, global governance has to go far beyond the regional level. European climate protection doesn't make sense, and global rules are needed to reduce hazardous volatilities of currencies. Global governance is an affair that takes many decades to develop. The strengthening of the United Nations is the recurrent melody of this task.

Civil society can play and does play an increasingly important role in these global affairs. NGO's can cry alarm whenever something scandalous goes on. Civil society consists of churches, trade unions, scientific groupings, philanthropic clubs and the whole gamut of charitable or at least not-for-profit NGO's. In many cases have these groups made themselves heard and forced the private sector to withdraw from unacceptable practices. In some case NGOs can cooperate with private companies on safeguarding particular public goods. A case in point is the "Marine Stewardship Council" formed by Unilever and WWF to ensure sustainable fishery by the firm. Figure 14.3 symbolizes the balancing role of civil society.

Without NGO power both national democracies and international treaties such as the Climate treaty or the Universal Declaration of Human Rights can easily be marginalized or ignored!

We have a long way to go both on national and international levels to develop a world society that is rooted in democratic control, in citizens' participation and in international fairness. But then, when Charles de Montesquieu developed his fundamental ideas in the 1740s about the need for the division of power, *no* division of power existed in his own country, France. It took a couple of decades until, as the first country in the world, the USA took up his ideas and created a democracy built on the division of powers.

Chapter 15
Limits to Privatization

Ernst Ulrich von Weizsäcker, Oran Young and Matthias Finger

15.1 Introduction: Seeking a Balance

Beware of extremes![1] What is needed is a proper balance: between freedom and order, between innovation and continuity, and—the topic of this book—between the private sector and the public domain.

Different actors will have their own preferences about the best balance between each pair of extremes. Poorer people may favour more order and a bigger public domain, while affluent people may want more freedom and more private ownership. Many believe private ownership promotes efficiency and wealth creation. Yet even affluent countries committed to the productive value of private ownership seek a balance between the private sector and the public domain, regardless of the tone of the rhetoric they employ.

Striking a good balance between private and public is the theme of this book. Thus we address one of the great issues of our times. Which services, including health, education, and welfare systems as well as law enforcement, infrastructure, and finance, can be entrusted safely to the private sector? Are there good reasons to keep a sizeable proportion of a society's land and natural resources within the public domain? Under what circumstances is it better for the state to own and operate (some of) the means of production? Where functions are performed by private companies, what sorts of rules and regulations should govern them?

There are no simple answers to these questions. Yet some important observations about recent developments are coming into focus. The thesis we develop in this book is that the recent and continuing swing toward privatization is in danger of becoming too much of a good thing. It may carry us beyond reasonable limits

---

[1] This text was first published in: Ernst Ulrich von Weizsäcker, Oran Young, Mathias Finger (Eds.): *Limits to Privatization—How to avoid too much of a good thing—A Report to the Club of Rome* (London: Earthscan, 2005). ISBN: 1-84407-177 4. Permission to republish this text was granted by Ms. Adele Parker, Rights Manager, Taylor & Francis Royalties Department, Cheriton House, North Way, Andover, Hampshire, UK on 21 March 2014. The book may be purchased at: http://www.taylorandfrancis.com/books/details/9781844073399/.

and produce undesirable consequences that outweigh the undeniable benefits of many forms of privatization.

In the body of this book, we describe and explore privatization in a range of sectors. Thousands of cases exist worldwide. Some have been successful by any reasonable standard, but others have failed. We have collected and selected, to the best of our ability, examples from around the world and from all relevant sectors.

What goes wrong attracts the attention of the media much more than what goes well. Never mind. It is widely assumed that privatization is a good thing; there is more need to understand its limitations and hazards than to confirm the assumptions underlying the trend.

At the time when our plans emerged to write this book, in 2001, we knew of no systematic worldwide study of privatization across all sectors. In the meantime, the World Bank has published a highly useful assessment of privatization of infrastructure in developing and Eastern European countries (Kessides 2004). This policy research report is unusually candid about failures, although it maintains the Bank's general assertion that privatization enhances efficiency, enlarges access to essential services and helps solve many development problems.

Before assessing actual experience through a series of short case studies and snapshots, we need to flesh out what we mean when speaking of limits to privatization and suggest strategies for rectifying the growing imbalance that has emerged in this realm. The following sections of this introductory chapter set the stage for our investigation.

15.2 A Word on Definitions

Like most prominent policy concepts, 'privatization' has a cluster of overlapping meanings. In this book, we use the term in its widest sense to refer to all initiatives designed to increase the role of private enterprises in using society's resources and producing goods and services by reducing or restricting the role that governments or public authorities play in such matters. This is often carried out by a transfer of property or property rights, partial or total, from public to private ownership. But it can be done also by arranging for governments to purchase goods and services from private suppliers or by turning over the use or financing of assets or delivery of services to private actors through licenses, permits, franchises, leases, or concession contracts, even when ownership remains legally in public hands. There are even cases such as 'build-operate-transfer' contracts where the private sector creates an asset, operates it for a certain length of time, and then transfers it into public ownership.

Where necessary, we use more technical definitions in accordance with the relevant literature.

Privatization is not the same as deregulation, which means the removal or attenuation of restrictions, including requirements and prohibitions, imposed by a public authority on the actions of public or private actors or, in essence, any reduction of state control over the activities of societal actors. Privatization often comes with deregulation, and especially the removal of exclusive rights and the opening up of a service to competition. But either can occur without the other. A major theme running through the book is that the success or failure of privatization is often strongly influenced by the kind and degree of regulation accompanying it. We draw a distinction as well between privatization/deregulation and liberalization, which means actions by governments to stimulate competition among companies in markets. Liberalization can take a variety of forms ranging from antitrust measures to the elimination of subsidies and the introduction of incentives to stimulate competition between private actors. Although they are separate phenomena, privatization/deregulation and liberalization often go together. In this book, we focus on privatization/deregulation, referring to liberalization as it affects privatization.

15.3 Changing Role of the State

Throughout much of the 19th and 20th centuries, governments assumed responsibility for an ever-expanding range of societal functions, in particular social and health functions. Later in the 19th century, but especially in the beginning of the 20th century, governments also played a crucial role in supplying *infrastructure*, such as highways, airports, port facilities, telecommunication and postal services, water supply and reservoirs, sewerage and irrigation systems, hospitals, schools, and increasingly in maintaining these facilities. Some infrastructure was actually built and initially owned by private entrepreneurs and later nationalized, notably many railways, power grids and telecommunication networks.

Nationalization was one of the central programmatic ideas of *communism* from the days of the Russian revolution onward. After World War II, communism spread further, and state-owned infrastructure *and industries* became the normal state of affairs throughout the communist bloc. But nationalization took place also in many western countries as well, notably in Europe and Latin America. Ideology (often socialist rather than communist) played a part. But nationalization was often motivated by pragmatic assessments that capital could be raised more readily and cheaply by the state, or services delivered in a more efficient, coordinated and effective way, if managed in the public interest as coherent and integrated systems rather than left to a patchwork of private companies who would inevitably put their individual commercial interests first. It is sobering to note how similar these arguments for nationalization—accessing investment, service responsiveness and efficiency—are to some of those advanced 30–40 years later to justify privatizing the same services.

States emerging from colonial rule during the 1960s and 1970s saw ownership and responsibility for essential infrastructure as the most natural task of the state.

By the end of the 1970s most of the world's governments had assumed far-reaching responsibilities for their infrastructure and various means of production.

In recent decades, however, the tide has begun to flow in the opposite direction. Since the 1980s, preceded by economic theory generated in the 1960s but significantly pushed by globalization, technological developments, market expansion, and the growth of private operators, a new trend has emerged of questioning the state's dominant role and of strengthening the private sector. This trend has manifested itself mainly in the privatization of an array of activities previously included in the public domain together with deregulation in the form of opening previously protected industries to competition and of reducing restrictions on the activities of private enterprises.

During the 1960s, the spirit of privatization went against established cycles of change and took on an air of courageous endeavor. Economists like v. Hayek (1944), Friedman (1962), and Coase (1974) paved the way. They treated the then fashionable search for social equity as inefficient, partly dishonest, and demoralizing for high achievers and thus, in the end, counterproductive to its own aims.

Neo-conservative politicians embraced their ideas as a basis for attacking affirmative action and other equity measures in the United States, Britain and elsewhere. Socialist and equity concerns were seen as a form of romanticism or a cover-up for state cronyism, if not as a dangerous ideological support for the arch-enemy, the Soviet Union.

In political battles to promote the emerging neo-liberal and neo-conservative paradigms, the most widely used term was "economic efficiency". The efficiency code was used successfully to de-legitimate 'inefficient' state functions in an ever growing array of sectors. The ideological adversary at the time was Keynesianism. Neo-conservatives said that Keynesian deficit spending was only leading to or deepening 'stagflation', the unpleasant combination of high inflation rates and high unemployment that was characteristic of the late 1970s. A lean and efficient state, by contrast, would promote tax reductions, low inflation, and better services—and that from the private sector (also see Bruno/Sachs 1985). The enlarged private sector, not the state, would take care of economic growth and technological progress. In retrospect, nobody would seriously deny that the paradigm shift has been successful by its own standards of efficiency.

But can we afford to assume that these arguments will hold for all cases? Or are there limits to privatization in the sense of thresholds beyond which the costs of privatization outweigh the costs?

15.4 Forms of Privatization

In this book, we devote particular attention to four broad categories of privatization:

1. putting state monopolies into competition with private or other public enterprises,

2. outsourcing, in which governments pay private actors to provide public goods and services,
3. private financing in exchange for delegated management arrangements, often with a view to transferring ownership to the state after a period of profitable use; and
4. transfers of publicly owned assets into private hands.

The first category includes cases in which governments alter the rules of the game to allow private enterprises to in specific activities. This might result in competition between public and private operators. In order to match private sector efficiency, the public operator often takes efficiency as a measure of performance. Examples include the admission of private airlines to compete with national airlines, the authorization of private universities and hospitals, and the deregulation of the electricity market.

The second category includes all those cases in which governments contract with private enterprises to supply specified goods and services previously provided by public agencies. Arrangements of this kind have a long history, going back, in the case of France, to the late 19th century. But recent decades have witnessed an explosion of new arrangements of this sort dealing with services ranging from education through health care, public transportation, security systems and on to prisons.

The third category involves the financing of infrastructure but increasingly also other projects (e.g., schools, hospitals, museums) by the private sector. These include Build-Operate-Transfer (BOT) arrangements, concessions and lease-contracts. The World Bank has been especially active in developing sophisticated models whereby the private sector provides the money in exchange for various forms of delegated management arrangements so that the investments can be paid back. Originally only practiced in developing countries, this approach to financing public investments is increasingly common in industrialized countries.

Perhaps the most familiar of these four categories is the transfer of publicly owned assets into private hands. Historically, governments have often sold public land to private owners or granted more limited use rights such as the right to harvest standing timber on public lands. More recently, we have witnessed a rising tide of sales of all or part of publicly-owned enterprises, such as national airlines, telecommunication companies, petroleum companies, and other energy companies.

Sketching out the four categories, we have mentioned quite arrange of legal forms. Figure 15.1 is meant to help distinguish the most important ones of them.

Government activities are coloured green, private activities red. In all cases, governments may impose more or less stringent regulations restricting the actions of private enterprises. Land may be transferred with easements, allowing public access for specified purposes (e.g. hiking, berry-picking). Government contracts with private corporations may or may not include requirements about service standards, access, prices, auditing, reporting and such like. Authorization to enter

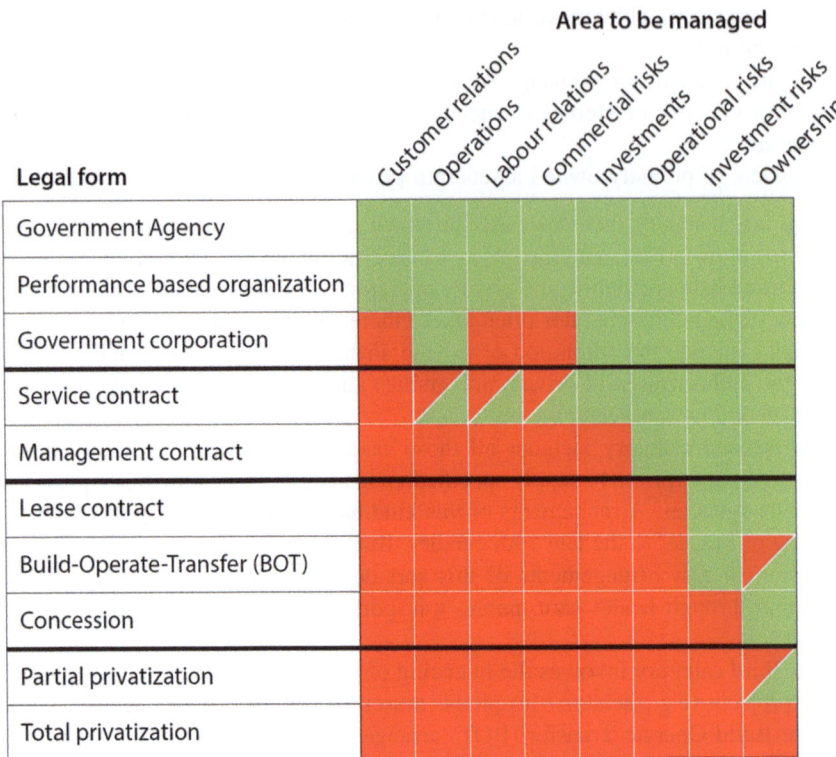

Fig. 15.1 Schematic overview illustrating different mixes of public and private activities

new markets may or may not include regulations dealing with occupational safety and health or with the treatment of pollutants.

15.5 Points of Departure: Polar Perspectives

Historically, arguments about privatization have often been polarized between extreme statist and extreme libertarian positions. The statist view, held by many socialists through the 20th century, is that it is desirable for the state to own property and to provide public services, and moreover that state agencies can be trusted to act in the public interest without needing any external regulation or scrutiny. The libertarian view, exemplified by Milton Friedman, calls not only for transferring property rights from the public domain into private hands, but also for minimising regulations and restrictions on the actions of owners of private property.

15 Limits to Privatization

Statists assume that the public sector will automatically operate in the public interest because there is no profit motive to make it do anything else; libertarians believe that the mechanism of competition driven by profit seeking will guarantee that private providers will respond to what users of services want—because those who fail to do so will be driven out of business by those who succeed. Thus these two views, in one sense opposite, share the assumption that the right *ownership* of public assets is the key to good results.

Both contrast with views that effective regulation is the key to ensure that services act in the public interest. The problems experienced with unregulated or weakly regulated private ownership have led many to conclude that private ownership can bring benefits if, but only if, there is a strong framework of regulation to ensure that companies look after the needs of all their customers and support broader public policy goals such as environmental protection and equal opportunities. 'Privatization—but with strong regulation' is perhaps the most popular position at the time of writing, and many of the examples in the book provide evidence supporting it.

However, arguments for regulation do not apply only to private ownership. Some now argue for public ownership, but with strong independent regulation to ensure that public services operate in the interests of their users, rather than of their staff, and to keep up the pressure to improve efficiency and effectiveness which is otherwise absent from services which face neither competitors nor shareholders. Perhaps this kind of regulation could have prevented the complacency, inefficiency and unresponsiveness in many public services that made privatization seem attractive or even essential. Some of the examples in this book are of initiatives to make public service providers more accountable, responsive and self critical in order to improve performance without resort to privatization.

The two axes of public versus private ownership and weak versus strong regulation thus imply four possible combinations, as follows (Fig. 15.2):

Of course, this is a simplification. As we have noted already, there are many different models of ownership, and even reducing them to a continuum from public to private is a simplification. Likewise there are many more aspects of variation in regulation that just a line from 'strong' to 'weak'. Moreover all four positions assume that governments are stable and capable of regulating and enforcing property rights. This assumption seems increasingly unrealistic in many societies for reasons having nothing to do directly with privatization.

The diagram should make one point clear. Privatization is not about doing away with regulation as some followers of Milton Friedman seem to believe. To the contrary, as de Soto (2000) argues, private ownership may require particularly reliable regulation to protect weak and small against the steamrolling powers of the rich. So this diagram should not be treated as more than a rough summary. Nevertheless, these first-order distinctions are helpful in drawing attention to the twin dangers of market failure and government failure. They may also help us to avoid confusing the ideal and the actual or, in other words, comparing ideal forms of privatization with the gritty realities of public supply or vice versa.

Level of regulation	Ownership	
	Public	Private
Minimal	Most public services before1980 (all in former soviet states)	Most 'classic' privatizations through 1980s
Strong	Recent experiments as alternative to privatization	More recent privatizations

Fig. 15.2 Four possible combinations of ownership and regulation

15.6 'Horses for Courses'

Different forms of privatization fit different societal and cultural conditions. Forms of privatization that produce positive results under some conditions fail miserably under other circumstances. In some settings, privatization has even given rise to corrupt practices and a tendency to defraud the general public. More generally, there are *limits to privatization* in the sense of boundaries beyond which the negative consequences or costs of privatization outweigh any benefits.

Thus, the central concerns of this book are:

- What are the limits to privatization in the sense of boundaries beyond which the net results typically turn negative?
- What conditions determine the location of these limits in particular issue areas and in specific social/political systems?
- What systems of regulation are needed to avoid or minimize both market and government failures?
- How can the best balance be achieved in allocating social functions or tasks between the private sector and the public domain?

15.7 Generic Pros and Cons

The consequences of privatization, deregulation, and liberalization are context specific. Identical structures or property rights or regulatory systems can and often do produce divergent results in different social settings. Many of those who promoted privatization in the wake of the collapse of the Soviet Union learned this lesson the hard way. For the victims that lesson was even harder! It is always important, therefore, to evaluate the consequences of privatization or liberalization in context rather than judging specific situations on the basis of broad generalizations.

Even so, we can identify the major categories of arguments advanced by advocates of privatization and liberalization as well as the core concerns of those who oppose these trends.

Proponents of privatization advance three broad types of arguments in one form or another:

- Privatization promotes efficiency and enhances social welfare by creating incentives to allocate resources to their highest and best use and by encouraging individual owners to invest in longer-term initiatives in the expectation that they will reap the resultant gains over time.
- Transferring property from the public domain to the private sector and reducing regulatory restrictions increases personal freedom, avoids the effects of rigid bureaucracies, and reduces corruption and cronyism in public places.
- A combination of private property and appropriate incentives and rules produces equitable results in the sense of rewarding those who work hard, take risks, and exercise ingenuity. As such, it also leads to creativity and innovation.

For their part, opponents of privatization advance a parallel set of propositions:

- Privatization tends to weaken the state and its capacity to care for social equity. By weakening the state, privatization also erodes the significance of democratic participation at national and sub-national levels.
- Privatization subordinates broader public goods, including long-term ecological and cultural values, to commercial imperatives.
- The need of private providers to make a commercial return (in the form of profits, dividends, rents and/or interest) adds to the cost of providing public services.
- Commercially optimal decisions are often suboptimal for public goals; competition forces providers to ignore externalities.
- The private sector never really does take over risks in public service provision. Where costs exceed revenues, private operators respond by demanding subsidies, raising charges, cutting necessary investment and maintenance, or walking away.

These considerations ought to be treated as a checklist of queries in terms of which to interrogate specific situations involving calls for privatization and liberalization or their reversal. The results of such evaluative efforts will vary from one functional area to another as well as from one social setting to another. But the use of such a checklist provides some assurance that the results will be based on a systematic assessment of the pros and cons of specific proposals.

15.8 The Shape of Things to Come

We proceed by presenting a set of short case studies and even shorter snapshots designed to cover the spectrum from successful instances of privatization to outright failures evaluated in terms of a range of criteria.

Some cases simply examine the trend toward privatization, including its early motives, driving forces, and consequences. These cases illustrate where the trend began—in what countries and in which functional sectors. Other cases show initial enthusiasm yielding to increasing frustration. Still others feature initial skepticism giving way to satisfaction on all sides.

In many cases—increasingly visible today—the costs of privatization and liberalization outweigh the benefits. Public protests, blatant corruption in the process of selling or transferring public assets, and clear breaches of the basic commitments are all visible signs of this failure.

Each of our case studies and snapshots seek to answer at least *one* of the following questions:

- Has the *quality of services* improved or deteriorated as a result of privatization; have existing services disappeared or new services been created or added?
- Has privatization altered incentives in such a way as to improve *efficiency or cost effectiveness?*
- Has privatization led to increasing or decreasing *investments in infrastructure?*
- Has competition *been increased through the process of privatization?*
- Has the pursuit of efficiency led to *distributive consequences* that are manifestly unfair or inequitable?
- Has privatization/deregulation generated *collective-action problems* in the sense of situations in which individual rationality leads to social outcomes undesirable for all? If so, have these problems been handled in a cooperative and fair manner?
- Has privatization/deregulation affected the supply of *public goods*, such as the fulfillment of basic human needs or the protection of the planet's life support systems?
- Has privatization/deregulation affected rates of *(un)employment* or otherwise impacted basic *social welfare?*
- *How was the* process of privatization (*bidding-process, contracting, monitoring* etc.) *organized? Has proper* regulation *been installed?*
- In what ways has privatization/liberalization affected the role of *democratic controls and democratic participation?*
- Has privatization/liberalization undermined or augmented *cultural diversity?*
- *Are* gender *effects visible or expected?*
- Has privatization/liberalization encouraged *capacity building,* in such forms as training individuals and exposing small and medium-sized companies to the discipline of international markets?
- Has privatization/liberalization increased or decreased the effects of *corruption* in the public domain or in the private sector?

- How has *globalization* affected the results arising from privatization/liberalization within *individual societies?*
- Has the *local economy* been included into the process or have local firms profited from technological transfer and capacity building?

Different case studies and snapshots illuminate different questions. We have presented them as readable narratives or vignettes rather than attempting to force them into a standardised format. We have not been able to pursue all the intriguing questions they raise. Nevertheless, we believe that our efforts have produced some important insights regarding limits to privatization. Understanding and evaluating privatization as one of the mega-trends of our time must be a 'work in progress'. We invite readers to arrive at their own assessments of experiences with privatization.

In reality, what we found it that some cases of privatization went truly well, but in the majority of cases, there were major downsides. What at first sight looks as "economic efficiency" can often be a massive deterioration of service quality. In the absence of strong state supervision, the private owners don't show sufficient interest in doing their job well and at publicly acceptable price levels.[2]

References

Bruno, Michael; Sachs Jeffrey D., 1985: *Economics of Worldwide Stagflation* (Cambridge, Mass.: Hardard University Press).
Coase, Ronald, 1974: "The Lighthouse in Economics", in: *Journal of Law and Economics*, 17,2: 357–376.
De Soto, Hernando, 2000: *The Mystery of Capital: Why Capitalism Triumphs in the West and Fails Everywhere Else* (New York: Basic Books).
Friedman, Milton, 1962: *Capitalism and Freedom* (Chicago: University of Chicago Press).
Hayek, Friedrich von, 1944: *The Road to Serfdom* (Chicago: University of Chicago Press).
Kessides, Ioannis N., 2004: *Reforming Infrastructure. Privatization, Regulation, and Competition* (Washington, D.C./New York: World Bank & Oxford University Press).

---

[2] This last paragraph of the Introduction is in fact a very condensed summary of the concluding chapter of the book.

Christine and Ernst Ulrich von Weizsäcker. *Source* Personal photo collection of the author

Chapter 16
Information, Evolution, and 'Error-Friendliness'

Ernst Ulrich von Weizsäcker and Christine von Weizsäcker

Abstract Information can be conceived as being composed of two complementary components: novelty and confirmation. Whenever either of the two is zero, information is zero. Genetic information, too, requires both novelty and confirmation. Evolution can be seen as the history of diversification. Selection alone reduces diversity. Recessivity appears to serve as a mechanism to protect diversity against selection. So does the geographical and behavioral 'separation' of species. Both recessivity and separation can be seen as 'error-friendly', a broader concept that is supportive of diversity, learning and further evolution. This concept should also be obeyed in technological applications.

16.1 Shannon's Information

What is it that travels from an observed object, whether it be prey, a darling's face or a sequence of black-and-white stripes an a rotating cylinder, then arrives at the retina, further continues to the brain and finally produces motoric or emotional reactions? Is it the same thing all along? Can Shannon's (1948) formula, which states that information is the negative logarithm of an event's probability, help answer the puzzling questions obstinately asked by Hassenstein (1966)? Or does it simply confuse and blur and prematurely simplify the philosophical question about the nature of the 'thing'?

We wish to thank Bernhard Hassenstein for many an inspiring dialogues, the reviewers of this paper for many valuable comments, and Robert Grant for linguistic editing and substantive comments. The present paper is based on a manuscript originally published in: Biological Cybernetics 79, 501–506 (1998) and dedicated to Bernhard Hassenstein on the occasion of his 75th birthday (Berlin-Heidelberg: Springer-Verlag). This text was published by MIND* matter, An International Interdisciplinary Journal of Mind-Matter Research, 4,2 (2006): 235–247. The permission to republish this text was granted on 25 February 2014 by the editor Harald Atmanspacher.

Shannon's co-worker Weaver (1949) seems to have supported the latter idea when he stated that: "Two messages, one heavily loaded with meaning, and the other pure nonsense, can be equivalent as regards information". Indeed, if information is defined the very way Shannon/Weaver (1949)—and, almost simultaneously, Wiener (1948)—defined it, Weaver's astounding statement is correct. Therefore, it can be concluded that the logarithmic definition of information is completely separated from meaning and will not help answering any semantic puzzle.

This confusion was not really cleared up but rather intensified when Brillouin (1956) took a closer look at some of the questions addressed in Shannon's and Weaver's book. Brillouin observed that the term information in almost all contexts is used to remove uncertainty and entropy, while the two terms information and entropy are used as being essentially interchangeable by Shannon. Shannon's choice of the letter H for entropy (from the Greek capital letter 'eta') was a direct reference to Boltzmann's thermodynamic entropy, though Shannon considered it to be a mere analogy. Brillouin went from analogy to homology and said that informational entropy and thermodynamic entropy were clearly identical. He therefore concluded that Shannon had used the wrong sign for information and introduced the new term 'negentropy' for information as removing entropy. Nearly everybody was pleased, and Brillouin's term negentropy made its way into the everyday language of intellectuals.

However, later the negentropy concept became unsettled by Zucker (1974). He noted that the removal of entropy for the creation of information presupposes the existence of a larger amount of entropy in the first place. If there is no entropy, there can be no negentropy either. He concluded that, in a sense, Shannon was quite right in using the same sign for both information and entropy. So, Zucker's contribution leads us back to square one with regard to Weaver's 'nonsense puzzle'.

16.2 Novelty and Confirmation

In most mathematical contexts, this whole debate could be disregarded insofar as everybody was able to quantify pieces of information regardless of the sign by using Shannon's formula. Philosophically, however, it remained a rather unsatisfactory state of affairs. The authors suspected that the debate about the sign was an indicator of a basic conceptual deficiency regarding the meaning of information (Weizsäcker/Weizsäcker 1972). We noticed that meaningful information, i.e., information defined in a way which is consistent with normal use and understanding of the term by biologists and other people, does not lend itself as being quantified by one single mathematical expression, whether it be entropy or negentropy. Both make sense but both are insufficient.

We proposed that meaningful information consists of two mutually complementary components, novelty and confirmation. This dual concept links up to the previous controversy in that novelty relates to entropy, confirmation to negentropy However, while entropy and negentropy by definition are mutually commensurate with one another, novelty and confirmation are not. They are both independent variables, so they cannot be plotted on the same axis and a second axis is required. Also, we observe that the mutual relationship between novelty and confirmation has a resemblance to a concept that in theoretical physics is called complementarity.

Any novelty, unexpectedness and surprise have to be embedded in a large envelope of confirmation, otherwise it cannot be understood, it cannot be meaningful. The process of understanding is more than the passive final destination for the transport of a message. It is, in fact, a very active process and involves the creation of new information. If understanding means any degree of learning on the part of the receiver, understanding is not confirmation alone: it also means adding novelty to the receiver.

Confirmation is a more complex and abstract concept than novelty. Confirmation comprises a variety of stabilizing factors such as (1) the channel through which a message is being transmitted (2) the code or alphabet in which a message is encoded, and (3) the body of knowledge the receiver possesses and may use to recognize and interpret an incoming message. Without a channel, the receiver is not able to identify a message as a message. Without a code, the receiver cannot decipher the incoming characters or signals. Without some degree of recognition and reassurance, the receiver is likely to be confused and unable to make use of the novelty obtained.

Shannon, of course, assumed tacitly the existence of a channel, of a reliable code, of a large body of receiver knowledge and of some degree of semantic recognizability of the message. If an appropriate channel structure, code reliability and pre-existing knowledge are present, novelty will appear as useful information, at least if that novelty comes packed in a meaningful message. If, and only if, all these preconditions are fulfilled, do we end up in the agreeable, straightforward situation where—as common sense tells us—novelty can be read to mean information and where it can be measured mathematically by the logarithmic entropy formula.

However, neither biological systems nor human communication are confined to areas where the incoming novelty is always perfectly attuned with confirmation at all semantic levels. Novelty arrives through a multitude of imperfect channels, mostly unplanned, scantily coded and full of noise. Biologists are not surprised to learn that when novelty grows too large, it is far from being useful and fully understandable. Some of the novelty may even serve to blur the code (e.g., by the appearance of unknown signs or words) or to endanger the functions of the channel. Such novelty cannot be valued as information. At best it is nonsense, at worst it engenders destruction.

Our term novelty is indeed closely related to what Shannon called entropy. But we have to realize that Shannon's definition of information as entropy only applies under a set of highly demanding 'agreeable' conditions. According to our terminology, these agreeable conditions are those with a large and comfortable excess of confirmation.

The amount of novelty is zero when everything is known and certain (probability p—1) and it grows as matters become uncertain ($0 < p < 1$). This means that it follows the mathematical behavior of the negative logarithm of probability It grows with shrinking probability. At probability p—0, the negative logarithm is infinite. Total novelty, therefore, is physically impossible. This is also plausible from an ontological point of view.

Both confirmation and novelty may be quite different for different receivers, which again is not surprising according to the common meaning and experience with information. Meaningful information is "information as if the receiver mattered", an allusion to a famous saying by Schumacher (1973) who pursued "economics as if people mattered". A message can be novel and incomprehensible to one receiver but may be perfectly understood by another receiver. It may even be fully understood by the first one who may be first puzzled, then learning and finally grasping the message after several repetitions (confirmations) in a similar context. Here we are at the core of animal and human learning behavior. Learning invariably goes with repetition.

Shannon's formula requires a high degree of confirmation. In fact his formula is built on the complete stability of the alphabet used, of the expectation probabilities and of the channel. This means that outside the highly stable, agreeable, excessively confirmed situation, the formula is invalid because values of expectation probabilities cannot be reliably established.

If we turn to the level of letters of the alphabet, stable probabilities can be established by looking at the statistical properties of randomly selected texts. Even for short combinations of two or perhaps three letters, the statistical properties may still be found. However, statistics becomes more difficult as the combinations grow longer, i.e., if we are talking about the statistics of entire words. We are confronted with the phenomenon of context dependence. The word 'Shannon', for example, is likely to appear either in texts covering information theory or in texts on Irish geography, but it is very unlikely to appear anywhere else.

For words it is at least possible to think and talk about probabilities. For entire sentences, the very notion of statistical probability is difficult to retain. At this level of complexity and contextualization, the degree of novelty is simply too high for the application of Shannon's formula. Significantly, however, it is sentences that carry meaning.

The high degree of confirmation at the low level of letters or phonemes is very useful for the deciphering of noise-affected messages. If the letter combination 'Shannon' is slightly distorted to 'Skannon' in a text about Irish geography, everybody will, without much ado, consider the letter k as 'noise' and correct it.

Even if the sender may have the very unusual intention of introducing a new letter to the alphabet (e.g. 'Shaennon', to emphasize a certain type of pronunciation) the receiver will at first ignore such novelty. The sender has to repeat and emphasize his unusual (novel) intention to make himself understood. He has to wrap his novelty in confirmation.

The high redundance of the alphabet and, less so, of the word dictionary, is not only useful for correcting noise-caused errors. It also serves as the precondition (the necessary excess confirmation) for communicating novelty at the higher levels of sentences, entire messages, articles and books.

Even if both the novelty and confirmation preconditions are fulfilled, we still do not have a guarantee for the usefulness of the information conveyed. We cannot discuss usefulness without taking a closer look at the receiver. As Weizsäcker (1971: 352) put it, information is only what is understood (by the receiver). In the same spirit, he also says, information is what creates new information. On the one hand, a nonsense message does not create new information and will not be understood. On the other hand, determinism, meaning total confirmation, has no place in the context of biological information either, because it excludes the core issues of understanding and contextualization.

Biological communications at their best follow patterns where new information leads to understanding and understanding opens new information. They always include an element of novelty, be it the novelty of the situation or of the errors in transmission. A very desirable kind of novelty appears in learning. It can be seen as new comprehension, i.e., new confirmation available for future information events, on the part of the receiver. Taking into account novelty and confirmation in their essential contributions to information, one could say: information is that which is almost understood.

16.3 Genetic Information

Our considerations on information, novelty and confirmation in the field of abstract and linguistic structures could be valid also in the context of genetic information and the genetic code. This is plausible because one finds a stable alphabet with stable probabilities of the four letters occurring on normal DNA. These letters are adenosine (A), cytosine (C), guanine (G) and thymine (T).[1] To be more precise, there are four *base pairs*, A–T, C–G, G–C and T–A, on a double-stranded DNA helix. However, the partners of each pair mutually determine each other so that the base sequence of each single strand determines the sequence of

---

[1] Rarely, some letters other than A, C, G and T occur, such as hypoxanthin (HX), but in the replication process they will be treated as if they were one of the four standard letters.

the opposite strand; the opposite strand represents highly useful confirmation, except for rarely occurring copying errors. On the single-stranded messenger RNA alphabet, the letter T is absent and is functionally replaced by uracil (U).

Four different DNA—or messenger RNA—letters permit 16 different sequences of two DNA letters and 64 different triplet sequences. It is the triplets that are used to code for one letter each (out of 20 allowed letters) at the next informational level. Twenty different amino acids are used at this level as 'letters' of the protein alphabet. This is the famous genetic code. Any alteration at the DNA level can lead to a change in the sequence of amino acids. Some redundancy (confirmation) is provided in the translation mechanism against errors: most amino acids correspond to two or even four different triplets of DNA/RNA letters. Hence a fair number of alterations (mutations) at DNA level do not lead to a change of the respective amino acids.

Beyond these basic and simple structures of the genetic code, there are more complex mechanisms of information transfer involved. Amongst these is the three-dimensional structuring process of proteins. The shape of the structure depends not only on the sequence of amino acids but on the production process in time and space. Different procedures may lead to different structures even if the sequence of amino acids remains identical. All this results in considerable unpredictability, flexibility, etc., i.e. novelty.

One basic mechanism certainly plays a prominent role in the development and use of genetic confirmation and novelty. It is the sheer duplication of DNA strands. The doubling (confirmation) of one strand later allows one of the two strands to experiment with unusual (novel) sequences on itself while the other remains unaffected and can be relied upon if the experiment fails. Ohno (1970) devoted a whole monograph to this phenomenon which appears to be fundamentally important for the trial and error processes of biological evolution. In the abstract language of the initial paragraphs of this paper: genetic information makes systematic use of confirmation to allow for experimentation with novelty.

Information is what creates new information also sounds like a general maxim for genetic information. DNA sequences which are not eventually transcribed into messenger RNA would seem to be of less value for life (although they may have an important role to play in stabilizing the DNA body). Messenger RNA, in order to become 'meaningful', has to interact with both the ribosomal structure and the availability of transfer RNA, including their attached amino acids.

Wherever information occurs in biological systems, be it at the level of nucleic acids, of antibodies, of travelling excitations in neurons or of complex cerebral 'messages', it is the very creation or transmission of new information that makes the information useful and triggers off cascades of differentiated informational events.

16.4 Evolution

The processes of biological evolution seem to follow the same pattern of living and creative information. Novelty is represented by mutations. Without mutations, evolution would stagnate. Confirmation, on the other hand, is related to fitness in the Darwinian sense.

Fitness depicts the fitting of a key with a lock. If the lock changes, the same key no longer fits. Thus, fitness must not stagnate. Fitness cannot be as narrowly defined as the scores of a soccer game or the speed of a sprinter at the Olympic Games. Many more than three 'champions' receive a medal of victory. The puzzle about Darwinian evolution is not the survival of the fittest but the exuberance of the survival of the less fit. This is the origin of the miraculous diversity of life forms.

When Charles Darwin, back in London from the Galapagos Islands, had a closer look at the birds' bodies he had brought from the islands, he observed a surprising number of new species of finches with very unusual adaptations. He concluded that finches arriving on the islands were able, in the absence of rivals from other taxa, to develop original and appropriate adaptations.

One phenomenon of diversity in particular needs to be recalled to our attention: the gene pool. Historians of biology know that Darwinism had a very difficult time in the 1920s when serious calculations were made which demonstrated that the occurrence of "hopeful monsters" was extremely unlikely. At that time, monstrous mutations like the bithorax Drosophila fruitflies were used as the paragon of mutations. Obviously, the likelihood for monsters to turn out beneficial was correctly calculated to be close to zero. It was not until the Neodarwinist school of Fisher (1930), Haldane (1932) and Wright (1931) established the concept of the gene pool that Darwinism was resurrected. The Neodarwinists essentially discarded evolutionary steps of monstrous size from their considerations and relied on a multitude of minute mutations which over millions of years were allowed to accumulate in the gene pool of a species. This was seen as the usual process of evolution leading eventually to miraculous achievements.

One remarkable thing about the gene pool is that it allows the 'less fit' genes to accumulate if they are recessive. A recessive gene is weak in expression in the phenotype. If the matching gene from the other parent remains 'wild type', the offspring's phenotype is wild type. This is the situation which geneticists call 'heteroallelic' (where 'allele' stands for a gene). Only if the recessive gene is inherited from both parents (the 'homoallelic' situation) does the offspring show the recessive gene's specific properties.

The mechanism of recessivity seems almost to have been designed to protect lots of less fit genes from being eradicated. This plainly contradicts the simplistic perception of the Darwinian selection mechanism. However, seen from a long-term evolutionary perspective, recessivity appears rather ingenious. Combined with the steady occurrence of new mutations, recessivity leads to the accumulation of a huge diversity of genes and their combinations in large populations. The result is ample variance but hardly any monsters. If the population is decimated by environmental stress, parasites or predators, small pockets of mutually separated populations are likely to survive. In these very small populations, inbreeding drastically increases. This enhances the probability for the homoallelic occurrence of recessive genes, and the once-hidden genes have a high likelihood of becoming visible. There will be a certain distinct probability that some of those genes will be useful for overcoming the new stress factors.

Goldschmidt (1940), Mayr (1942) and Gould (1980) (see also Eldredge/Cracraft 1980) were instrumental in establishing the evolutionary power of inbreeding and separation (or isolation, as the earlier authors called it). In this modern pattern of evolutionary theory, not only recessivity but also separation itself appears as an ingenious mechanism of defending and enhancing variance. Barriers, geographical and non- geographical, help protect local varieties and thereby preserve a larger diversity than would be possible if the Darwinian selection mechanism eradicated all the less fit variants everywhere and instantly.

Sometimes recessivity is depicted as a lamentable mechanism which prevents the eradication of unwanted, even lethal, genetic factors. However, such complaints do not do justice to the mechanism. Quite often, the recessive gene carries a feature that is useful for survival. Perhaps best known are the sickle cell and betathalassemia genes which give malaria resistance to heteroallelic carriers. There are indications that the heteroallelic carriers of the cystic fibrosis gene have an advantage in surviving infant diarrhea, and that the heteroallelic disposition for diabetes may be advantageous in times of hunger.

Simplistic Darwinism now appears as an absurdly static view of evolution. Such a narrow interpretation does not account for novel stress situations to which the existing wild type may not be well adapted compared to some sleeping, recessive variants.

Moreover, certain evolutionary achievements such as the size of the dinosaurs or the huge teeth of pre-neolithic sabre-tooth tigers do not guarantee success under all circumstances. Evolution must not be streamlined too much. It has to be 'permissive' and open to be successful in the long run. In addition, mutations should not be avoided or erased even in a very successful species. As Fisher put it, "the rate of increase in fitness of any organism at any time is equal to its genetic variance in fitness at that time" (Fisher 1930: 35).

Less fit variants are protected against premature selection, and a sufficient supply of new less fit variants is constantly provided by the mechanism of mutation. Mechanisms that ensure the survival of the less fit and the maintenance of diversity must be extremely old and fundamental to the working of evolution. Eigen/Schuster (1979), in their famous paper on the evolution of the prebiotic 'hypercycle', play a game with the symbolic sequence of letters TAKE ADVANTAGE OF MISTAKE.

16.5 Error-Friendliness

The protection and creation of weak variants seems to be a rather universal phenomenon, not restricted to biological evolution. It can be observed in all learning processes. To conceptualize and popularize the underlying common phenomenon, one of the authors (C. v. W.) proposed the new term error-friendliness 'Fehlerfreundlichkeit' in the late 1970s. Error-friendliness comprises the robustness to withstand internal or external attacks and errors but it goes beyond this function. It also includes the invitation and accommodation of variance, experimentation, novelty and even errors. However, such experimental errors should not as a rule be allowed to rock the boat. Novelty should be embedded in or matched by a sufficient amount of confirmation. The two authors, after long, sometimes heated but always cooperative discussion, later summarized the results in a joint scientific paper (Weizsäcker/Weizsäcker 1987).

Pedagogy always emphasizes the essential role of errors in human learning. A magnificent way of gaining experience with errors is playing. In the animal kingdom, playing is widespread, chiefly among young animals. However, adults also assume a crucial role in playing: they reduce the risks involved in the play and games of their offspring. They are part of the error-friendliness of the learning environment. The young produce the novelty and the errors, while the adults secure confirmation and come to the rescue when needed. Error-friendliness in playful and in serious situations means that novelty and confirmation are in balance and that risks are being kept at a manageable level. Letting children climb onto a sofa in the living room is tolerable and increases skill and judgment. Letting them climb onto the window sill of an open window of the fifth floor is not.

Plausible criteria for parenthood and education have their correspondence in technology. Construction errors are tolerable for a hut, but not for a skyscraper or a highway bridge. Skyscrapers are not very error-friendly from an architect's point of view. This does not mean they cannot or must not be built, but they ought to be built only if every single part of the design and the construction machinery is perfectly reliable. Whenever possible, the planning of skyscrapers ought to make use of the advantages of error-friendly learning processes and their results. As a matter of fact, the construction business is well aware of this. The architect can 'play' with his design on a computer before the first stone has been moved. He or she will always plan for a certain extra robustness of the construction to absorb unexpected challenges. Earth movers, concrete manufacturers, scaffold makers and all craftsmen are building on years and years of experience involving lots of errors suffered and later overcome.

The situation is somewhat trickier in complex systems such as the nuclear energy cycle. The nuclear energy cycle involves lots of situations where one would wish the erroneous experiences never would occur. We are not surprised by the fact that no country going nuclear dares to demand full private insurance coverage from its nuclear power installations, although that coverage is routinely demanded from and obtained for the chemical industry, the construction business or any other standard industry.

The concept of error-friendliness may give a clue to the mathematical puzzle why 10^{-6} 'mega-deaths' are not equivalent in our common sense assessment with one death. Anything involving an appreciable probability of mega-deaths is instinctively seen by ordinary people as not being error-friendly.

Unwanted mega-risk situations should be expected to occur also in the relatively new field of genetic engineering. In particular, the deliberate release into the environment of genetically modified organisms (GMOs) appears to involve problems that cannot be routinely handled by insurance companies. We do not know how ecosystems will react to the abundance of 'ferries' for genes or viruses crossing the borders of species, in view of the fact that the evolutionary 'invention' of species barriers proved so successful in balancing the co-evolution of hosting organisms and their pathogens.

Particular worries can be associated with the release of plants or microorganisms carrying all kinds of resistance factors against herbicides and antibiotics. Another disturbing event is the introduction into plants of genes producing the Bacillus Thuringiensis (BT) toxin. Entire crop fields of such GMOs create the evolutionary conditions for BT-resistant pests, which could mean the end to organic farming: organic and traditional farmers use BT if really needed as a convenient pest control not interfering with the food quality.

Unexpected harm to beneficial insects, like bees and lacewings, caused by genetically engineered crops, give scientific indications of the large dimension of the task of identifying and tracking the complex effects of such releases on ecosystems worldwide.

What is perhaps most worrisome in newly designed plants is the slowness with which the effects can be observed. Neophytes—exotic species of earlier centuries typically needed several generations as a lag phase to develop the vitality for spreading in their new environments. Why should GMOs behave differently? Who will keep the records and continue with the monitoring? Will innovations outpace learning and thus become nonsensical and destructive?

Genetic engineering in drugs and in environmental clean-up jobs may not be as harmless as the business community seems to think. A 1997 report by the National Institute of Health in the USA assessing benefits and risks concluded that the effects are not yet well understood and recommended a "back to the lab" strategy as a general attitude. Waste sludges from manufacturing may theoretically turn into sources of novel and potentially hazardous germs.

Generally, it can be said that the willingness of insurance corporations (including reinsurers) to cover the risks of a certain technology and the willingness of the producers of this technology to seek insurance contracts are fair indicators for the error-friendliness and sustainability of the respective technological pathways. Unfortunately, biotechnology industry still bound to short-term shareholder-value strategies fights efforts to introduce obligatory insurance coverage for the risks linked to genetic engineering technologies.[2]

Once the abstract concept of error-friendliness was created, it was easy to associate it with a large number of well-known concrete mechanisms. One ubiquitous mechanism of securing error-friendliness is compartmentalization. In evolutionary theory, separation was seen as a geographical fact, as a means to keep diversity high, and to prevent parasites and other infectious nuisances from affecting the entire world. In the body, the cellular structure provides:

- redundancy: if one cell fails, there is another one available to do the same job;
- cell differentiation: if one cell mutates to do a new job, there are others left to do the old one.

Taken together, these features make the cellular organization eminently error-friendly. Other concrete biological mechanisms of error-friendliness *include*:

- the immune system, allowing the body to cope with novel chemicals;
- mutations;
- recessivity (see above);
- curiosity, inducing individuals to explore their environment, to run risks and to learn; and
- aging and death, enabling a species to balance novelty and confirmation under the given constraints in space.

Bearing in mind the difficulties, errors and political atrocities caused by the use of scientific terms for the description of social phenomena, but being pressed by the constraints of time and space for this paper, the authors would like to invite readers to test the concept of error-friendliness and to reflect on its usefulness in

---

[2] The negotiations of the Cartagena Protocol on Biosafety (CPB) within the framework of the UN-Convention on Biological Diversity (CBD) were concluded in 2000. The Protocol contains legally binding agreements on the transboundary movement of "living transgenic organisms" (the CBD-term for genetically modified organisms) necessary for the protection of biodiversity and human health. It has come into force in 2003 and has been ratified by 134 countries (25/09/2006), see www.biodiv.org. The Precautionary Principle, non-subordination of environmental and health considerations, Advance Informed Agreements (AIA), clear identification and international liability are key issues of the Cartagena Protocol on Biosafety. Unfortunately, none of the main exporters of genetically modified organisms, USA, Canada and Argentina, has ratified as yet. And the implementation of liability at the international level (CPB, Article 27) has been postponed again and again.

social contexts. Some indicative hints may be permitted to show that civilizations have—without using the term.

- developed a large number of error-friendly mechanisms, including:
- democracy, allowing people to change government as long as the errors are small enough not to warrant the troubles of a revolution;
- division of power;
- freedom of press, allowing a free flow of information including embarrassing errors to those who can then alert the public and thereby initiate societal processes leading to remedies;
- overcoming dogmatism in the scientific endeavour: allowing errors and aberrant theories to come up and to test them using old and new methods (Mittelstrass 1997);
- accident insurance (see above).

All in all, error-friendliness can be seen as a new concept, as a new magnifying glass to perceive and describe the cybernetic functions of living systems, notably of those undergoing change.

References

Brillouin, L., 1956: *Science and Information Theory* (New York: Academic Press).
Eigen, M.; Schuster, P., 1979: *The Hypercycle* (Berlin: Springer).
Eldredge, N.; Cracraft, J., 1980: *Phylogenetic Patterns and the Evolutionary Process* (New York: Columbia University Press).
Fisher, R., 1930: *The Genetical Theory of Natural Selection* (Oxford: Oxford University Press).
Goldschmidt, R., 1940: *The Material Basis of Evolution* (New Haven: Yale University Press).
Gould, S.J., 1980: "Is a new and general theory of evolution emerging?", in: *Paleobiology*, 6,1: 119–130.
Haldane, C.B.S., 1932: *The Causes of Evolution* (London: Harper).
Hassenstein, B., 1966: *Kybernetik und biologische Forschung*, Akademische Verlagsgesellschaft Athenaion. Frankfurt. (Separate printing from Handbuch der Biologie Vol. 1/2).
Mayr, E., 1942: *Systematics and the Origin of Species* (New York: Columbia University Press). (Reprinted in *Evolution Now. A Century after Darwin*, ed. by Smith, J.M. Freeman, San Francisco 1982: 129–145).
Mittelstrass, J., 1997: Vom Nutzen des Irrtums in der Wissenschaft, in: *Naturwissenschaften*, 84: 291–299.
Ohno, S., 1970: *Evolution by Gene Duplication* (Berlin: Springer).
Schumacher, E.F., 1973: *Small is Beautiful. A Study of Economics As If People Mattered* (London: Blond & Briggs).
Shannon, C.E., 1948: "A mathematical theory of communication", in: *Bell System Technical Journalm*, 27: 379–423, 623–656.
Shannon, C.E.; Weaver, W., 1949: *The Mathematical Theory of Communication* (Urbana, IL: University of Illinois Press).
Weizsäcker, C.F. von, 1971: *Die Einheit der Natur* (München: Hanser).

Weizsäcker, E. von; Weizsäcker, C. von, 1972: „Wiederaufnahme der begrif-flichen Frage: Was ist Information?", in: *Nova Acta Leopoldina*, 37,1: 535–555.

Weizsäcker E. von; Weizsäcker C. von, 1987: "How to live with errors. On the evolutionary power of errors", in: *World Futures*, 23: 225–236. Originally published in German: Fehlerfreundlichkeit. In Offenheit - Zeitlichkeit - Kom-plexität, ed. by K. Kornwachs, Campus, Frankfurt 1987, pp. 167–201.

Weaver, W., 1949: "The mathematics of communication", in: *Scientific American*, 181,1: 11–15.

Wiener, N., 1948: *Cybernetics* (New York: Wiley).

Wright, S., 1931: "Evolution in Mendelian populations", in: *Genetics,* 16: 97–159.

Zucker, F.J., 1974: „Information, Entropie, Komplementarität und Zeit"., in: E. von Weizsäcker (Ed.): *Offene Systeme I. Beiträge zur Zeitstruktur von Information. Entropie und Evolution* (Stuttgart: Klett): 35–81.

Ernst Ulrich von Weizsäcker shaking hands with Chinese President Hu Jintao in 2007 during the State Visit of the German Federal President Dr. Horst Köhler in the Great Hall of the People in Beijing. *Source* Personal photo collection of the author

Chapter 17
Factor Five: Transforming the Global Economy Through 80 % Improvements in Resource Productivity

Ernst Ulrich von Weizsäcker, Karlson 'Charlie' Hargroves, Michael H. Smith, Cheryl Desha and Peter Stasinopoulos

17.1 Introduction: Factor 5—The Global Imperative

... The 21st century will see monumental change.[1] Either the human race will use its knowledge and skills and change the way it interacts with the environment, or the environment will change the way it interacts with its inhabitants. In the first case we would use the sophisticated understanding in areas such as physics, chemistry, engineering, biology, commerce, business and governance that we have accumulated in the last 1000 years to bring to bear on the challenge of dramatically reducing our pressure on the environment. The second case however is the opposite scenario. It will involve the decline of the planet's ecosystems until they reach thresholds where recovery is not possible. Following this we have no idea what happens next. If the average temperature of our planet's surface increases by 4–6 °C we will see staggering changes to our environment, rapidly rising sea level, withering crops, diminishing water reserves, drought, cyclones, floods. Allowing this to happen will be the failure of our species, and those that survive will have a deadly legacy.

In this book we support the many recent calls from leading governments to achieve 80 % reductions in environmental pressures and provide a reasonable and realistic approach to reaching this target by 2050, leading to and requiring already profound innovations across industries, across communities and across cultures.

The results of the 2006 *Stern Review* of the economics of climate change (Stern 2007) provide a glimpse of what is coming upon us regarding global warming. Business-as-Usual (BAU)—that is, a continuation of growth trends from the past without any serious decoupling of growth from carbon dioxide (CO_2) emissions—will lead to doubling annual emissions compared with the amounts

---

[1] This text was first published in: Ernst Ulrich von Weizsäcker, Charlie Hargroves, Michael H. Smith, Cheryl Desha, Peter Stasinopoulos: *Factor 5: Transforming the Global Economy through 80 % Improvements in Resource Productivity* (London: Routledge, Earthscan, 2009). ISBN: 978-1844075911. Permission to republish this text was granted by Ms. Adele Parker, Rights Manager, Taylor & Francis Royalties Department, Cheriton House, North Way, Andover, Hampshire, UK on 21 March 2014. The book may be purchased at: http://www.taylorandfrancis.com/books/details/9780415848602/.

from 2000. If we are able to stabilize CO_2 concentrations at 450 parts per million (ppm), we would have to reduce annual emissions by 50 % at least, which is a factor of 5 less than the BAU scenario. Even this extremely ambitious trajectory will not prevent some additional global warming in the range from 1 to 3.8 °C. Coming to grips with this situation may require more capacity than our communities possess. Responding to the challenges now faced may require more understanding than our professions possess. And admitting that change is needed, and needed fast, may require more humility and courage than our typical national leaders possess; legally speaking, they are accountable to their respective national constituencies, not to the Earth.

The purpose of this book is not to repeat the litany of problems that face us. This has been thoroughly and rigorously presented in a number of recent works by UNEP, OECD, the IPCC and individual authors such as Lester Brown, Al Gore and David Suzuki. Nor is the purpose of this book to depict economic growth as the inevitable reason for destruction, as has been the motto of Mishan (1967), and much of the *Limits to Growth* debate of the 1970s. Nor is it our purpose to decry capitalism as the ultimate evil, as has become fashionable in our days after the deep dive the world economy took since the second half of 2008. But we surely join critics of capitalism to a certain extent—as some features of deregulated financial markets have been disastrous and demand careful re-regulation.

What's more, in the context of the ecological state of the world, there is a need for regulation to prevent capital from investing in destructive industries and instead encouraging investment in value-creating activities conserving natural treasures. The purpose of this book is to inspire hope. It is not good enough simply to present a highly theoretical picture of how technology could save the world. Instead we want to present practical pictures of whole systems of technologies, infrastructures, legal rules, education and cultural habits interacting to produce economic progress while conserving a healthy environment. Virtually all the strategies outlined in this book can be applied now by nations, companies and households to achieve Factor Five. This 'whole system approach' will also help overcome the *rebound effect* of additional consumption gobbling up all technological efficiency gains that were meant to save resources and conserve the environment.

To fill this message with real world substance, we present numerous examples of resource productivity improvements from the most relevant sectors, showing that the said Factor Five, or 80 %, reduction of environmental impacts per unit of economic output, is available. This multifaceted universe of opportunities represents the core body of our book.

While we strongly advocate for significant improvements in resource productivity, we add, however, that there will also need to be consideration of aspects related to *sufficiency* (discussed in Chap. 11 of the original book). We shall need some rules of constraint or insights into other forms of satisfaction than the maximization of monetary throughput, or GDP. Relating to capitalism and regulation, we repeat and support my understanding from some 20 years ago that 'communism collapsed because it was not allowing prices to tell the economic truth, and that capitalism may also collapse if it does not allow prices to tell the ecological truth'. Markets are superb at steering an efficient allocation of resources and stimulating innovation, but they don't provide public order and law, moral standards, basic education and infrastructures, and markets are miserably inefficient, often even counterproductive, when it comes to protecting the commons and steering innovation into a long term sustainable direction. Human societies, and the environment, will need a healthy balance between public and private goods, or between the state and the markets, as suggested in Chap. 10. The mindset dominating much of the world during the past couple of decades of weakening and ridiculing the state, was gravely mistaken. We do need strong states and engaged citizens working together to create good legal and moral frames for the markets. Moreover, citizens, nation states and the international communities of states and of citizens are expected permanently to act on those markets, as consumers, innovators, workers and guardians against destruction, and for technological and civilizational progress in harmony with the conditions of nature.

Whether we want it or not, we are in the midst of highly political issues when getting serious about protecting and restoring the basis of life on Earth.

17.1.1 Balancing Economic Aspirations with Ecological Imperatives

Balancing private with public goods means to a large extent balancing economic aspirations with ecological imperatives, and during early human history there has been no visible evidence of such balancing. The environment seemed endless, for all practical purposes, and was often seen as hostile. Human civilization developed by taming wild parts of nature and harnessing the powers of nature to extract some of the natural treasures and resources. Economic survival and the increase of welfare seemed like the natural mission and mandate for humanity, with impacts on the environment remaining a negligible affair. Even during early industrialization, until as late as the 1960s, environmental impacts of economic activities looked mostly like local and peripheral concerns—a steel mill, a chemical plant, a textile dye factory here and there. A small power plant to supply the energy needed for factories caused local air and water pollution and local health problems, but the environment as such, outside the cities, was not seen as affected. It was not until the 1960s, when human population was approaching four billion people that 'the environment' became a major political issue.

Overcoming some initial resistance on the part of industry, democratic states such as Japan, the US and Canada, and the West European states initiated and adopted pollution control legislation, with muscles to enforce the law—and it worked. The cleaning of industry made rapid progress. Banning a few particularly unhealthy substances, filtering exhaust gases and purifying waste waters, and finally redesigning some processes were the means of decoupling industrial outputs from polluting nuisances. After a mere 25 years, the foam hills on rivers had

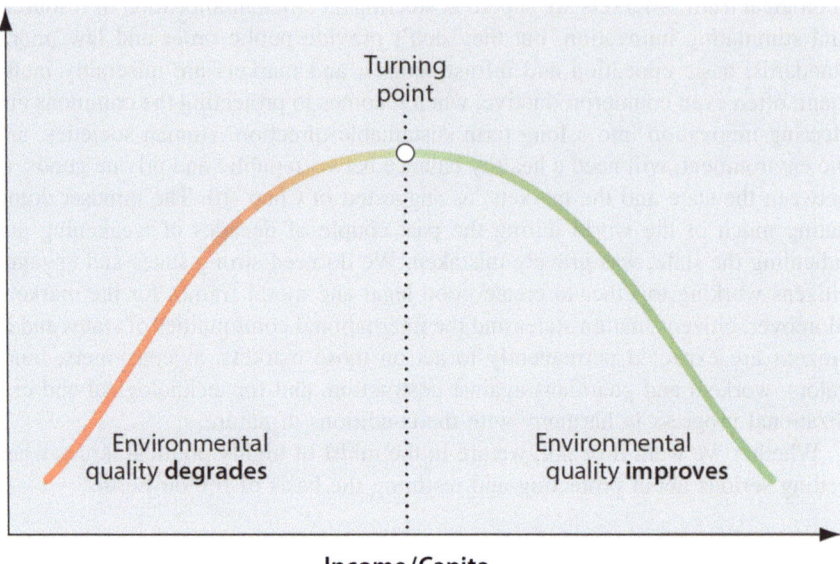

Fig. 17.1 The convenient paradigm of the *Kuznets Curve* of pollution. The historic development is typically from *left* to *right*, meaning that countries start poor and clean, then they industrialize to be come rich and polluted, until they are so rich that the can afford strict pollution control so that they end up rich and clean. The *Figure* was prepared by The Wuppertal Institute

disappeared and the industrial agglomerations, such as the Ruhr in Germany, Osaka in Japan, or Pittsburgh in the US got cleaner than they had been for a 100 years. This success story surprised many sceptics who had seen the cause of the problem as economic growth as such. The lesson from pollution control seemed even to reverse such earlier fears: it was the rich, and further growing, democratically organized countries, or regions inside countries, that were the most effective in cleaning their environment, leaving the dirt to the poor and to the non-democratic societies. A very attractive new paradigm emerged—the Kuznets curve of pollution, shown in Fig. 17.1, whereby countries having the economic maturity and financial means to deal with pollution control would engage in this agenda and move towards a wonderful harmony of 'rich and clean'.

During the 1980s, this convenient paradigm began to dominate the debate about the mutual relation between economic aspirations and ecological imperatives. It became perfectly respectable even for environmentalists to say: 'let the economy grow and take care of environmental concerns later'. Understandably, this attitude became the standard frame of mind of the leaders and representatives of developing countries in all international environmental negotiations. And the rich countries of the North had little to put against this view, as well as having no intention to contradict anyway.

Alas, the Kuznets curve paradigm does not work for the global environmental problems of our days. Pollution control is actually a very restricted part of environmental reality. Impacts such as climate change, resource shortages and biodiversity losses follow a completely different logic from pollution control. In reality, it is the 'rich and clean' countries that are the biggest cause of such impacts. Carbon footprints so far relentlessly grow with increasing prosperity. The situation gets much worse for the rich if historical carbon emissions are also counted. The 2004 per capita cumulated carbon dioxide load from the US was about 1000 metric tons, in China it about 60 metric tons, in India 25, in Germany nearly 800 (World Resources Institute (undated)). Figures get still somewhat worse for the rich countries if world-wide supply chains are considered as well. Many countries have outsourced the energy and carbon intensive segments of the supply chain to countries like China, which thereby got ever larger carbon footprints in the service of others.

In any case, the strong correlation between carbon footprints and GDP led many people in the US as well as in the developing countries to believe that reducing CO_2 emissions was tantamount to reducing economic welfare and was therefore politically unacceptable. The most convenient way of dealing with that situation was, of course, denying that there was any scientific proof of global warming or of human causes for additional warming. Fortunately, that's the past, at least as regards the attitude of the US government. But the challenge remains to find a new and healthy balance between economic aspirations and ecological imperatives.

Addressing global warming is the most prominent aspect in our days of ecological imperatives. But biodiversity protection is no smaller challenge. Biodiversity losses result mostly from land-use changes, and these usually occur in the service of more production, that is, more economic growth. The most commonly used measure for land-use in the service of consumption is 'ecological footprints', which estimates the land required for specific goods or services, or for the standard of living of average citizens of different countries. US (or European) footprints include land areas abroad that are needed for goods and services consumed at home. Bananas from Ecuador, copper from Chile, palm oil from Malaysia and the life cycle footprints caused by computer manufacturing in China, all count towards the European footprints to the extent that Europeans consume bananas, copper, palm oil or computers. It is therefore not surprising that ecological footprints tend to be biggest for the rich, as can be read from Fig. 17.2.

... What seems to be missing is a clear sense of direction. It was there in earlier centuries when prosperity growth with little regard to the environment was the guiding philosophy nearly everywhere. There were technological breakthroughs from time to time, spurring growth and creating a sense of excitement. Those breakthroughs included the steam engine, railways, electricity, cars, chemical technologies, radio and TV, and, most recently, IT, biotechnology and nanotechnology. Also the globalization of industrial supply chains can be seen as a breakthrough, notably in terms of keeping consumer prices down. However, there are signs of fatigue with this kind of progress as it hits its natural limits. For the

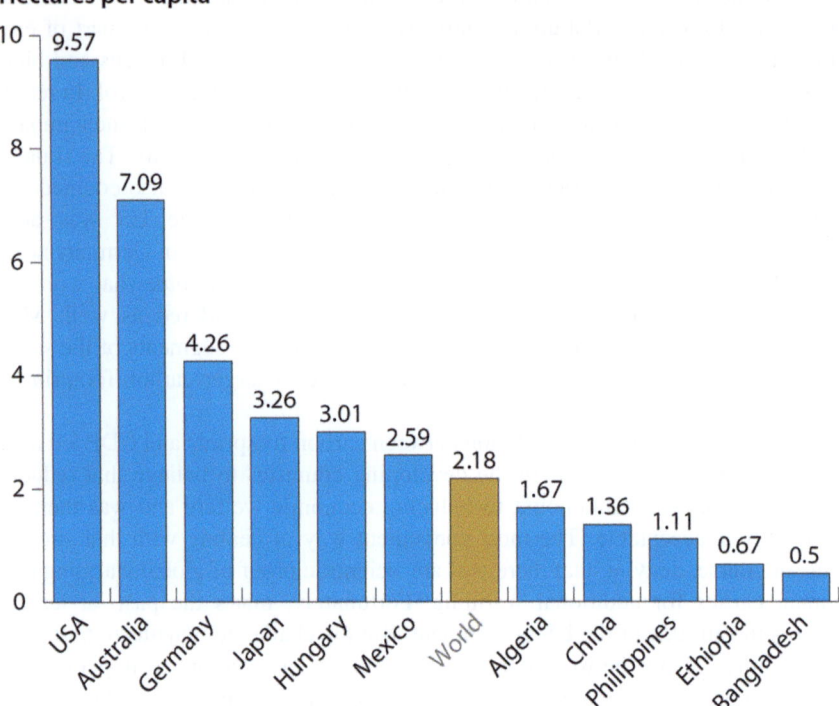

Fig. 17.2 Footprints of people in different countries. *Picture* Global Footprints Network

world economy to find its way back to healthy and robust development, a new and reliable sense of direction will be needed.

Providing this new sense of direction is the basic motive of our book. We suggest that we are at the dawn of a new long-term cycle, a new 'Kondratiev Cycle', or wave of innovation. During a time of recession, commentators often speak about, and hope for, the 'next upswing'. Usually it is the short kind of business cycles people have in their minds. But there are also long-term cycles, every 30–50 years, which can be attributed to major technological innovations, such as the ones mentioned above. Although standard economic literature does not necessarily accept the idea of long-term cycles, they have been a useful way of describing, characterizing and perhaps even explaining historical periods that are associated with technology driven major economic upswings. The best known early scholar to describe such long-term cycles was the great Russian economist Nikolai D. Kontratiev (1892–1938).[2] His pivotal book was called *The Major Economic Cycles* and was published in 1925. Kondratiev himself had no strong

---

[2] Jacob van Gelderen and Samuel de Wolff are the two Dutch economists who proposed long-term cycles as early as 1913, but their work, written in Dutch, remained unknown to N. Kondratiev, J. Schumpeter and others and was translated into English only recently.

emphasis on technological change, but Joseph Schumpeter, the famous Austrian and later American economist, saw business cycles and long-term cycles as associated with major technological innovations. It was Schumpeter himself who suggested honouring Kondratiev (killed in 1938 by Stalin's 'Purge' firing squads), by calling the long cycles 'Kondratiev Cycles'.

Our point is that, according to historical evidence since Kondratiev's pivotal work, the magic of technological innovations tends to fade after some 20–30 years of its beginning. So it may not be too surprising that even the most exciting recent wave of innovations in information technology, biotechnologies and, somewhat more recently, nanotechnologies, is no longer strong enough to support worldwide economic growth.

Fading excitement with certain technologies would not yet make for a massive—and sudden—economic downturn. The arrogance and failures of much of the financial sector was the obvious cause of the present crisis. But if we want the economy to gain strength again, an exciting new wave of technologies might be the biggest hope for the world. A couple of years before the present crisis, Paul Hawken, Amory Lovins and Hunter Lovins, in *Natural Capitalism*, also summarizing the theory of long-term cycles, came up with the suggestion of a new industrial revolution unfolding, with energy and resource efficiency at its core (Hawken et al. 1999). Building on from this pivotal work, Charlie Hargroves and Michael Smith from The Natural Edge Project, and co-authors of this book, suggested in their 2005 book, *The Natural Advantage of Nations*, that the emerging wave of green technologies could be seen as the beginning of a new Kondratiev Cycle and offered the following optimistic picture for it (Fig. 17.3).

As we have observed before, some greening of technologies and the green economy is already underway. We do suggest that the process of greening, being the logical answer to the environmental constraints, will generate the new and reliable sense of direction that could pull us out of the recession. For this to happen, some additional momentum will be highly desirable. If the conviction spreads that the greening trend is inevitable and can take the shape of a full-size Kondratiev Cycle, we are confident that the desired momentum will come. Investors then have clarity about where to put their bets (Fig. 17.4).

… Greening the economy is perhaps a popular way of characterizing the innovations we expect to happen in the course of the Green Kondratiev. But we suggest going one philosophical step further. We observe, as economic historians are likely to agree, that the first 200 years of modern age economic development had the 'increase of labour productivity' as the one unifying motto. Labour productivity rose at a pace of roughly 1 % per year during the 19th century until the middle of the 20th century. From then on, owing to the accelerated global spread of technologies, progress increased by about 2–3 % per year. Overall, labour productivity has increased 20-fold over those last 200 years. Figure 17.5 shows a time window of some 120 years marking the impressive acceleration after World War II.

Today, labour is not in short supply. Otherwise the International Labour Organization (ILO) would not speak of a shortfall of 800 million jobs to create a

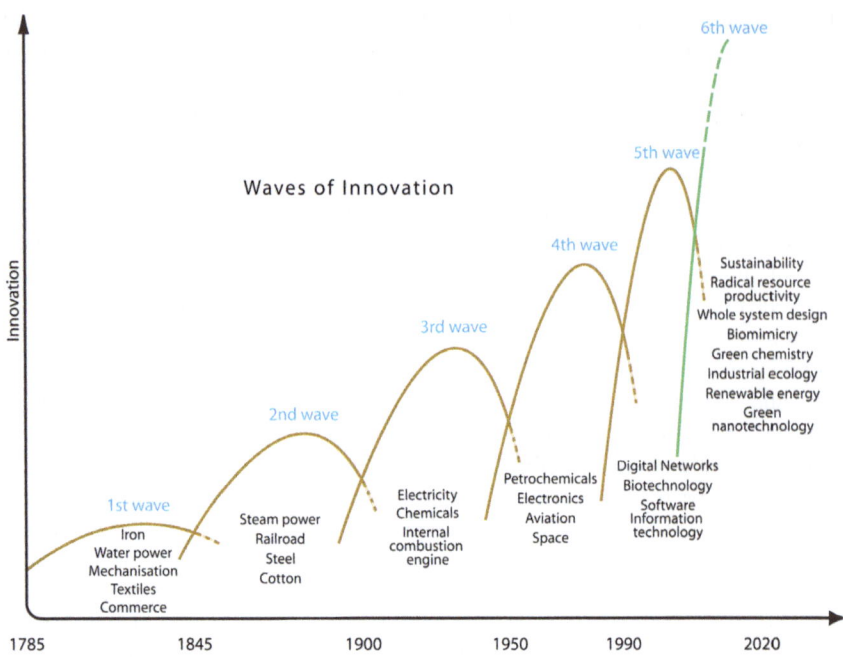

Fig. 17.3 Five classical "Kondratiev Cycles" may be followed by a *green* Cycle, characterized by sustainability technologies, including notably increased resource productivity. From Hargroves/Smith (2005)

situation of near full employment. On the other hand, as we have indicated before, energy and other natural resources are in short supply, and the scarcity is getting worse every decade.

This situation calls for a reversal of the emphasis on technological progress. Resource productivity should become the main feature of technological progress in our days. Countries making the scarce production factors more productive should enjoy major economic advantages over those ignoring the new scarcities. This is another way of emphasizing the need for a new technological cycle and a new orientation for the world economy, for national economies, and for individual firms. To relate this to the long cycle considerations, the green Kondratiev should become the first cycle during which resource productivity grows faster than labour productivity. In developing countries, the increase of labour productivity will, of course, remain a high priority because they want to catch up with industrialized countries. But they should avoid doing so at the expense of resource productivity. Many studies show that such a focus will help to boost the economy and create jobs, while reducing environmental pressures. As The Natural Edge Project explain in their upcoming publication *Cents and Sustainability*, investments in resource productivity transform and stimulate the economy in many ways (Smith/Hargroves 2010).

US-Dollar/h

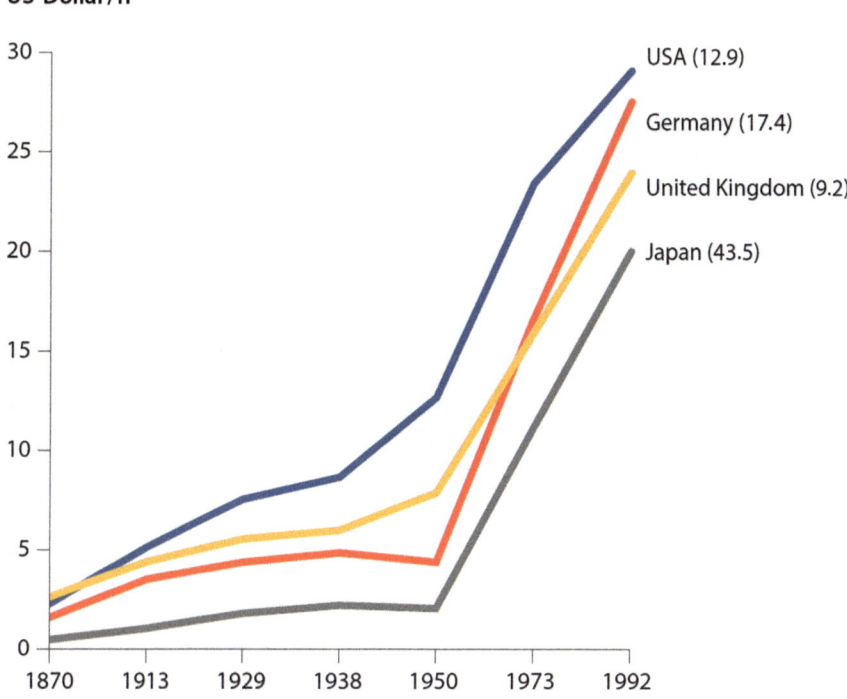

Fig. 17.4 Labour productivity has increased dramatically during the last century. *Figure* by Bleischwitz (1996)

... The next question is how the new Kondratiev can be kicked off. Do we have to wait for the whole world to be persuaded towards the new paradigm? Can single countries or companies go it alone or do they need a broadly accepted business environment for resource productivity and the rest? We suggest that pioneers can go ahead prudently with little if any risk on their economic performance. Philips has decided to concentrate on LED for example, and Toyota went ahead, together with Honda to introduce the hybrid car and was very successful domestically and abroad. Véolia Environnement pioneered 'city mining', the extraction of valuable metals from old waste dumps. Japan during the period from 1974 to 1980 went ahead with the phasing out of energy intensive manufacturing such as aluminium smelting from bauxite, and celebrated fabulous successes in other branches such as electronics and optics. Germany became the leader in renewable energies through a law of generous feed-in tariffs.

However, in the absence of certain framework conditions such as rising petrol prices, high electricity prices in Japan, high scrap metal prices from 2003 to 2007 and global warming concerns, it is less clear if such pioneers could have been successful. This is slightly different from the pioneers of the IT revolution who did not really need favourable framing conditions to make their technological advances a commercial success. This indicates that for the Green Kondratiev cycle to really take off, some political decisions on framing conditions would be most welcome.

As a matter of fact, we devote the entire second part of this book to the politics, economics and psychology of framing conditions for a massive launch of the exciting new efficiency and renewable energy technologies that are available right now. We describe them in the following Part I of this book. The transition into the Green Kondratiev cycle may actually be less dramatic than one might fear. The systems approach taken in this book essentially means redesigning the systems of industry, transport, buildings or agriculture, which were found to be destructive to the climate and the environment. The process of redesigning, as radical as it may be in terms of a new philosophy, can be a gradual and smooth one, encouraged by prudently designed and predictably changing framing conditions. We don't need to lose much physical or financial capital that is invested in our industrialized world. We only have to avoid investing fresh money into outdated and destructive operations and technologies.

17.2 A Long-Term Ecological Tax Reform

17.2.1 Introduction

Part I of this book focused on presenting a whole system approach to resource productivity based on best practices across a range of industries to demonstrate that Factor 5 improvements were not only possible, but often being achieved. The chapters adhere closely to an academic style, with comprehensive referencing and notes, to ensure a rigorous presentation. The innovation over that of the work presented in *Factor Four* lies in this whole system design approach to resource productivity gains—a transition that may be called 'transformational'—but it's still based on facts and figures. Chapters 6–8 then described a range of existing policies and instruments that are being used to underpin efforts to improve productivity. Chapter 8 ended with the observation, grim as it may be from an environmental point of view that improvements in efficiency in the past were almost always overwhelmed by additional overall consumption, be it by an increase of population, higher standards of living or just by consumer behaviour. Moreover, that in the real world, many of the exciting opportunities lying in the efficiency revolution remained mostly unrealized.

Amory Lovins' revolutionary 'Hypercar' has been around as a promising concept for over 15 years, but despite this the real revolution taking place in Detroit during that time went in the opposite direction—with the mass manufacturing of gas guzzling sports utility vehicles (SUVs). Retrofitting homes and commercial buildings to the 'passive house' standard is being driven by either research or progressive regulation in just a few European countries, with the odd standout around the world. On the whole the rapid uptake of plasma TV screens in OECD countries is offsetting many of the energy efficiency gains achieved. Construction of '6 star buildings' is happening only in signature buildings in cities around the world. Practical applications of light emitting diodes (LEDs) are still rare, present for the most part in specialty markets such as traffic lights and car brake lights. The cement industry is still indulging in the high energy intensity limestone based Portland cement and shows little intention of moving to alternatives such as 'geopolymers'. Water efficiency has remained a dream for most of the world's farmers with underground water reservoirs around the world being depleted at a mind boggling rate. And the idea of a 'cyclical society' is only slowly moving from its conceptual stage to reality in Japan and a few other Asian countries. Hence, such efforts are being overwhelmed by the shear increase in consumption. The rebound effect, described in Chap. 8, is the dominant reality worldwide, attenuated a bit since 2008 by the unwelcome advent of the economic crisis.

On the other hand, as presented in the Introduction, the ecological crisis is real and extremely dangerous. The world simply cannot afford to ignore it. The innocent phrase 'sustainable development' taken seriously implies truly transformational changes in all countries. To fill this postulate with substance, we suggest that a fivefold increase of resource productivity (at least) becomes a full-size reality in most of the world's countries, and that the rich countries finally learn to live with a degree of modesty. Two matters of course characterize reality everywhere in the world: (1) consumers consume what is affordable to them, and (2) producers produce what is profitable to them. Neither should be blamed for that. If we want the transition to a sustainable society, we must respect these matters of course and make use of them. It could look like this:

- Resource productivity should become ever more profitable for producers.
- Resource saving goods and services should be more affordable to consumers than resource wasting ones.
- While life supporting goods and services should remain affordable, high consumption of natural goods should become increasingly more expensive.

Conventional environmental policies have hardly touched these dimensions of sustainable development. This has to change, and it is going to be a transformational change. But it will be unavoidable if this world is to maintain its beauty and the support systems for human life. This last part of our book is devoted to the political and cultural dimensions of such transformational changes. One additional and contentious topic on the agenda of sustainable development is population increase. It lies outside resource productivity and we cannot deal with it in the context of this book, but we recognise its crucial importance and emphasize that political and religious leaders should acknowledge population as one of the central problems of our days.

How can we make resource productivity ever more profitable for producers and resource saving goods and services more affordable to consumers than resource wasting ones? The evident answer, in a market based global economy, lies in the price of resources. This chapter advocates a strategy of focusing on resource pricing through tax reform in a consistent and highly predictable way over the medium to long term.

Two different methods have been identified in Chap. 7 for pricing the relevant resources, notably carbon, energy and water. The two methods, which have been described as almost equivalent, are tradable permits and resource or emission taxes. Tradable permits in theory allow fixing the target levels of consumption and letting prices adjust to supply and demand changes. A trading regime in theory would be very efficient. On the other hand, taxes allow fixing prices (within limits) and leave the reaching of the target more or less undecided. This comparison seems to speak for tradable permits as the preferred instrument. However, at the end of Chap. 7, we realized that carbon markets as they are discussed in the real world of climate negotiations are far from leading to the reductions needed to stabilize climate. Such reductions would require very low caps and worldwide equitable commitments, both of which are far from the current reality in 2009. The previous chapters also mentioned that price fluctuations that may serve to adjust to oversupply or undersupply of permits are in danger of inviting speculation and causing major problems to investors and to socially vulnerable people. Nevertheless, carbon markets could well be the only feasible instrument at the international level, allowing countries a continuous exchange of permits. Also it was pointed out that environmental taxes can have a strong steering effect and can be designed such that capital destruction and social hardship are avoided. Revenue neutral green taxes reducing the fiscal load on human labour have been proven even to have net positive effects on employment and the economy. It may be that a long term, revenue neutral ecological tax reform is most likely to instigate both producers and consumers to make strategic and permanent efforts towards energy efficiency.

This chapter elaborates on this idea and discusses some of the questions that are expected to be asked by critics. Before going into detail, the chapter investigates what effect markets have had on primary resource prices over the last two centuries. The chapter focuses its attention on resource prices for two reasons, namely that they are one of the crucial factors determining the profitability of resource productivity, and secondly that they are the main target of a long term ecological tax reform.

17.2.2 Two Centuries of Falling Resource Prices

Surprisingly, primary resource prices (which also include primary energy) have had a tendency to fall over the last 200 years, as can be seen in Fig. 17.5. Knowing that prices are the strongest and most pervasive language in a market economy, this period of falling resource prices has induced industry and consumers to

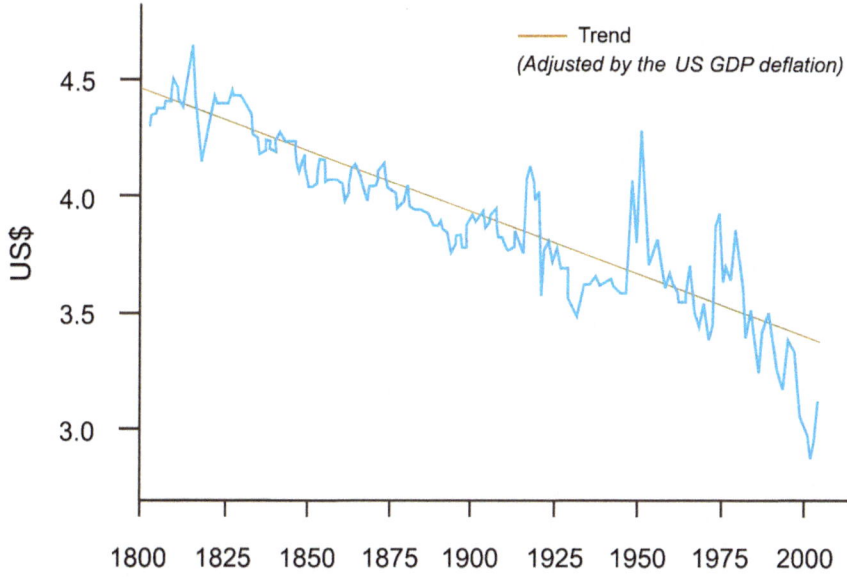

Fig. 17.5 During two centuries, resource and energy prices had a tendency to decline. (Real raw industrial commodity prices, inflation adjusted, in US terms, 1800–2004). *Picture* by TNEP based on data of The Bank Credit Analyst, 2004

become ever more negligent about resource efficiency. Exceptional phases such as the world wars and the energy price hikes of 1973–1982 and 2000–2008 have sent signals of scarcity into the economies of the world, but the basic assumption remained that mineral and energy resources are simply 'available'.

Prospecting, mining and transport technologies and related industries were the main drivers of bringing prices down.

The price hikes between 2000 and 2004 are indicated as an upward movement at the far right end of the curve. Some additional price increases occurred between 2004 and 2008. That recent trend brought prices back into the lower confidence interval of a secular downward trend. But when the economic crisis hit in late 2008, commodity prices tumbled back to essentially the lowest levels ever measured.

Before continuing the discussion of which instrument to apply and how to tailor it, let us look at the political landscape that has allowed resource prices to fall. It has been the explicit objective of business leaders and politicians for centuries to make natural resources available at ever lower prices. For business, lower resource prices simply improve the bottom line, and no one can afford paying a higher price than competitors do. And for politicians, it is extremely convenient, making people and businesses enjoy 'free lunches'. This political intention found its extreme expression in the Soviet Union, which put no price on environmental resources such as air and water, promulgating that all natural resources belonged to the public. A national policy that supported the use of free resources did not encourage conservation by individuals or firms. The logical result in the Soviet Union, aside from creating a sense of communist justice among the poor, was an extremely wasteful kind of industry. Soviet citizens wasted energy, water and minerals to an extreme degree, certainly contributing to the ultimate collapse of the Soviet economies.

Of course, political preferences alone don't reduce resource prices. The long-term downward trend of resource prices was also supported by technological progress in exploration, mining and processing. And market competition served to reward low cost suppliers. The Earth's geology has not so far shown any serious signs of depletion, except for the limited fossil fuels. Prospecting and mining was always systematically stepped up whenever commodity prices went up. Also, countries, companies and consumers became more modest or more energy efficient during these periods. This was the case during and after the two World Wars and after the two more recent oil price shocks of the 1970s and since 2000. Reduced energy intensity, trained during the 1970s and after, actually served as a key factor in stabilizing industrialized economies some 15 years later, when another quadrupling took place from US$9/barrel in 1998 to US$35/barrel in 2000. The recent price hikes after 2000 are usually explained by soaring petrol imports from China and India. This was indeed the reason for an imbalance between supplies and demand and justified prices rising to roughly US$100 a barrel.

However, the last upwards push between 2005 and early 2008 sending prices through the roof was actually triggered mostly by quite reasonable considerations on the part of institutional investors. Noticing that commodities were playing an increasing role in financial markets they thought they should perhaps diversify their assets and include holdings of raw commodities. But then, even a modest 1 % of their assets allocated to commodities meant a huge and unprecedented additional inflow of funds into oil and other resources markets. Some speculators, realizing the mechanics of this new development, joined the booming market driving oil prices near the staggering level of US$150 a barrel. It was not until then that some very oil dependent economies, most visibly the US, were pushed into a recession. Houses financed with sub-prime mortgages lost much of their value after commuting became a financial nightmare to the new homeowners, and that was the beginning of the sub-prime mortgage crisis and the financial meltdown. When the economies of the world began to shrink, oil and other commodity prices also collapsed from their speculative heights, reverting back to the long-term trend shown in Fig. 17.5.

This description is surely at variance with the mainstream of the 'peak oil' discussion that started in 1956 with M. King Hubbert's anticipation of oil production to peak around 1970. After the discovery since the mid 1970s of major additional oil reserves, the peak oil discussion faded off until the first years of this century. Also some other mineral resources such as copper, indium (used in transparent electrodes), rhodium (used in car catalytic converters) or phosphates (used in agriculture) were now included, as they are in high industrial demand that cannot be met with sufficient supply at present prices. Annual discovery rates of these resources have been falling behind annual consumption rates for some time.

The peak oil discussion has certainly helped oil producers and traders defend rising oil prices. It was also important for alerting the broad public to certain real limits to resource consumption. However, aside from the lower demand and lower prices during times of sluggish economies, we don't expect peak oil and gas to be as dramatic in the early parts of the 21st century as it has been predicted by some authors, including Heinberg (2007). The reason is simple—coal supplies have a geological reach of some additional 200 years, and coal liquefaction, using classical Fischer–Tropsch technology, would allow mass production of oil at prices of US$60–80 per barrel, assuming even high coal prices at US$100 per short ton.

The conclusion in this section is that markets driven by supply and demand are very unlikely to tell the full story of long term scarcity. And when scarcity becomes apparent or when the public can be made believe that the end is near, prices have a tendency to skyrocket. Both low prices and skyrocketing prices send wrong signals. The former invites squandering, the latter causes social and financial disruptions. We suggest that an enlightened leadership can do a lot to smooth out such fluctuations and can elegantly steer technology and behaviour into a high price world and highest efficiency.

Ernst Ulrich von Weizsäcker together with Prof. Ye Ruqiu (*centre*), as Co-Chairs of the Task Force for economic instruments for Energy efficiency and the environment of the China Council for International Cooperation for Environment and Development (CCICED). During the work of this Task Force, the ideas of a long term ecological tax reform (Sect. 17.2 of this volume) were formulated. On the *left* is Mr. Kai Schlegelmilch, Vice Chair of Green Budget Germany, on the *right* Mr. Ren Yong, then serving the CCICED, and Stefan Bundscherer of the German Society for International Cooperation, GIZ, Beijing

17.2.3 The Concept of a Long-Term Ecological Tax Reform

The last remarks above lead directly to the core of our proposal. We propose a long term policy decision of actively increasing resource prices. Our central idea is to create a broad political consensus on a trajectory of steadily progressing energy and commodity prices (inflation corrected). The main criteria for such active price policy should be:

- simplicity, also in terms of administration and control;
- predictability and reliability;
- social and economic acceptability.

In effect, based on these criteria, a trajectory of energy and other primary resource prices rising in proportion with resource productivity increases is evident. Then, by definition, there would be no social hardship on average, and hardly any capital destruction, as those who achieve the average increase will feel no impact, those who exceed it will benefit, with only those who fall behind the average efficiency improvement feeling the economic pressure. The socio–economic effect of this strategy can be expected to be positive for economies such as China, Bangladesh, the US, Japan, Germany and Egypt—all being countries importing energy and raw materials. They will be gently induced to become less dependent on such imports, leaving more domestic capital and wealth for distribution. Some of the commodity exporting countries and companies will conversely suffer export losses. But they represent a minority of countries, and a predictable and smooth transition will allow them to adjust and to move themselves into an economy of high added value and low carbon consumption.

Which is the best method of making raw material and energy prices rise predictably and in steps corresponding to their respective productivity gains? A cap and trade regime may qualify, while gradually reducing cap levels. On the international scale, as said before, cap and trade may be the only regime that is working. This is because it is very difficult to imagine countries agreeing on a common tax level for carbon emissions or fuels as well as on the distribution of revenues. As also said before, cap and trade regimes, with their unavoidable fluctuations, have a tendency of frustrating investors and hitting the poorest strata of society more than the affluent.

A political decision of fixing prices many years in advance sounds like a much more attractive method. In countries such as China, energy prices have been fixed by the state for decades. These countries can theoretically decide on the suggested price trajectory without much debate or a need to justify their decision against the fiction of a 'free market'. Other countries have the option of taxing energy and other primary raw materials. Our proposal is that this is done by adopting legislation requiring annual adjustments in response to measured resource productivity gains. To avoid too many debates each year, the adjustments can be foreseen every 5 years only, which coincides with China's regular 5 Year Plans. As different sectors develop their efficiencies differently, a political preference is possible for different price trajectories for those different sectors. However, in macroeconomic terms the best strategy would be to have the price signal uniform, thereby letting some sectors grow and others (slowly) shrink.

The idea of linking energy and other resource prices to gains in resource productivity has two different roots. One is the social and economic policy consideration that resource prices rising in proportion with resource productivity will not on average cause any social hardship or unbearable cost increases with industry. The other is the observation outlined below (subchapter on the paradigm of a 20-fold increase of labour productivity) of gross wages having increased in parallel with labour productivity gains over more than a 100 years. This parallel development of labour productivity and gross wages has been the backbone of the Industrial Revolution and has surely benefitted essentially all people, in some way. Recognizing the logic behind this parallel development, investors have always exploited any given opportunity of increasing labour productivity, knowing there was hardly any safer bet for the future. However, at a time of high unemployment and scarce natural resources, progress by increasing labour productivity has become less rational from a macroeconomic point of view. Moving investors to give a higher priority to resource productivity would make today's societies more prosperous and more competitive.

Clearly the best way of moving investors in the new direction of steeply rising resource productivity would be a long term and predictable trajectory of rising resource prices.

17.2.4 Overcoming the Dilemma of Short-Term Instruments

Most of these old and new approaches of pricing energy and the environment have one problem in common—they tend to be measures that are introduced at once. Hence they are subject to the debate over whether the thing being taxed is well defined, and if the size of the tax is adequate for the intended purpose. It is essentially the question asked already by Arthur Pigou. These questions lead to the typical dilemma of short term (tax) instruments—either the tax is high and will hit polluters and consumers hard, creating problems for the poor and capital destruction for industry, or it is low and has very limited effects. There have been some good examples of overcoming this dilemma. The German waste water charges, mentioned in Chap. 7, were announced a few years before being actually collected and had their strongest steering effect during the announcement period when the charge was still zero.

Another success story for an ecological tax was the Swedish NO_x charge of roughly US\$4 per kilogram nitrous oxides emitted from power plants, with all revenues redistributed to the electric power sector. The redistribution system made sure that the sector as a whole did not suffer, but a strong incentive was maintained to stop polluting. As pointed out in Chap. 8, even in Sweden, which is arguably the country with the broadest experience and the strongest emphasis on green taxation, only 5 % of total tax revenues were green taxes. This may indicate that there is some room for a higher percentage.

Two other European schemes were introduced in small steps: (1) the British 'escalator' on transport fuels, introduced in 1993, and (2) the German ETR, introduced in 1999 (see Fig. 7.2). Both schemes were progressive, meaning that year by year the duty increased by small amounts. But the announcement of further steps had a major effect on customer behaviour, not surprising for psychologists. One of the major effects was that families would buy a more fuel-efficient car when the old one was taken out of duty. Railways and other public transport enjoyed a renaissance, and unnecessary trips were reduced. After decades of ever increasing fuel consumption, the ETR, together with the increasing world oil prices, made an unexpected turn around in the transport policy and the respective CO_2 emissions of these countries. The 'escalator' idea seems to show a way out of the dilemma of short term pricing instruments. Figure 7.2 in Chap. 7 compared the British and German experiences with Canada and the US. In these two countries, the increasing efficiency of compact cars was more than compensated (in the absence of a price signal), by the introduction of tax privileged SUVs and by added miles driven. No signs of recovery can be seen in the mostly outdated and inefficient railway system. This is the type of consumer and policy rebound to efficiency that we must try to avoid through the use of such instruments as the ETR.

17.2.5 The Poor, the Blue-Collar Workers, the Investors, and the Fiscal Conservatives

Objections against ETR or energy related taxes tend to come from both the left and the right—from advocates of the poor and blue-collar workers, as well as from the investor community and fiscal conservatives. Advocates of the poor talk of the relatively high importance of energy costs for the poor in terms of total purchasing power, with energy and water taxes said to be 'regressive', hitting the poor more than the rich. This assertion is based on the fact that taxing negative externalities usually entails exerting a burden on consumption, and since the poor spend a higher percentage of their income on consumption than the rich, any shift towards consumption taxes can be regressive. However, conventional regulatory approaches can affect prices in much the same way, while lacking the revenue-recycling potential of ecological taxes. Hence, one of the strengths of the ecological tax approach is precisely that, unlike regulations, it provides revenue for ensuring low-income groups are not worse off. This can be done by reductions in (regressive) consumption taxes, or increasing welfare payments such as providing additional payment to the unemployed, pensioners and the disabled.

This regressive effect is almost universally observed for water charges and taxes, and the removal of subsidies for drinking water (although generous water subsidies for rich farmers abound and tell a different story). For energy, the statement is true only for the wealthy countries because in the poor countries the rich tend to have more energy intensive lifestyles than the poor; however, a high energy tax can surely work against poor households' aspirations to reach more convenient and prosperous lifestyles. Hence, these statements about regressive effects of taxes can be countered simply by granting a tax free minimum tableau of, say, one gigajoule of energy per person per week, or 200 L of water per person per week. Then the really poor would actually benefit, while the burden would shift towards the middle income and rich strata of the society. This has been effectively applied in the water tariff system in Setubal, Portugal.

Blue-collar workers, too, have a tendency of disliking ecological taxes as they associate environmental regulation and taxes with job losses. They typically use similar lines of arguments with the additional apprehension that energy taxes would destroy industrial jobs. There is surely some truth in this fear—if industry is unable or not willing to change technologies and practices to reduce energy and water intensity. Part I of this book is taking much of the legitimacy out these fears, arguing that exciting technological options exist to make a range of sectors significantly more energy and water efficient, and thereby much less vulnerable to resource taxation, even to the point that it can become a lucrative ongoing competitive advantage, especially with slow moving competitors.

Nevertheless fears of job losses from environmental taxes or regulations are common. A nationwide poll, conducted in the US found that 33 % of those polled felt themselves 'likely' or 'somewhat likely' to lose their job as a consequence of environmental regulation (Goldstein 1999). However, a study undertaken as part of this poll found that virtually all economists who have studied the jobs–environment debate over the last 30 years agree that these fears are unfounded. In reality, at the economy wide level, the study concludes that there has simply been no trade off between jobs and the environment, stating, at the local level, in sharp contrast to the conventional wisdom, layoffs from environmental protection have been very, very small. Even in the most extreme cases, such as protection of forests, or closing down fisheries, or steps to address acid rain, job losses from environmental protection have been minute compared to more garden-variety layoff events. In practice the real economy-wide effects of much of the environmental regulation enacted has been to shift jobs without increasing the overall level of unemployment. Globally there are now significant numbers of people who work in the 'environmental industry sector' as a result of these regulatory changes. In fact, regulation-induced plant closings and layoffs are very rare, with the study showing that in the US, about 1,000,000 workers are laid off each year due to factors such as import competition, shifts in demand or corporate downsizing, compared to annual layoffs in manufacturing due to environmental regulation in the order of 100–3000 per year.

Further, there is significant evidence to suggest that applying ecological taxes while reducing payroll tax can help create significantly higher employment. In 1994, DRI and other consultancies commissioned by the European Commission modelled a scenario where all the revenues from pollution taxes were used to reduce employer's non-wage labour costs, such as social security payments, superfund payments and payroll tax. The study showed that employment in the UK would be increased by 2.2 million through such tax shifting.

Aside from such reassurances regarding jobs, there are also those that take a more radical view saying that the classical model of heavy industry and mass manufacturing is part of the ecological problem of our days and should be overcome by less consumption and a move to the service sector. Even the unions are becoming remarkably open minded about the need for structural change, as is documented in the European Trade Union Confederation's joint statement with the European Environmental Bureau (EEB) (European Trade Union Confederrration (ETUC)/EEB 2005). To smooth the transition, certain temporary exemptions for the manufacturing sector can be granted, as in Germany, to avoid the destruction of invested capital. Or the Swedish NO_x model can be applied, recycling all revenues into the sector so that the sector as such is not losing out (Millock et al. 2004).

Investors have a tendency to indiscriminately despise taxes. However, they need not fear a revenue neutral energy tax escalator. On the contrary, what they should fear is unpredictability and unfair national rates creating competitive disadvantages. If it is possible politically to reach a consensus about a long-term trajectory for green taxes, that could result in a 'heaven on Earth' for many investors. It would mean that they can confidently move into ambitious technological and infrastructural projects with very limited risks, thanks to predictability of one of the most important factors, plus efficiency gains, and the prospects of new markets. This will eventually lead to major advantages over competitors currently working under conditions of substantially lower resource prices, and therefore giving too little attention to the actual scarcity of these unrealistically priced resources. Just imagine a new oil price shock or a crisis of water availability, you would expect those companies and countries weathering the storm easily, which have developed a highly efficient way of using those scarce resources.

This leaves the camp of fiscal conservatives as adversaries of an ecological tax reform. Their views may be ideologically fixed, but they should acknowledge that green taxes have a tendency of falling over time, as energy or water consumption fall. The opposite is the case with income taxes, which have a tendency of increasing over time. For energy and other resource taxes, you would indeed need an escalator to avoid dwindling fiscal revenue.

References

Bleischwitz, Raimund, 1996: *Ressourcenproduktivität. Innovationen für Umwelt und Beschäftigung* (Berlin-Heidelberg: Springer).
European Trade Union Confederration (ETUC); EEB, 2005: *Make Lisbon Work for Sustainable Development*; at: www.etuc.org/a/982 4 May 2009.
Goldstein, E., 1999: *The Trade-Off Myth. Facts and Fiction About Jobs and the Environment* (Washington, D.C.: Island Press).
Hargroves, Karlson, Charlie; Michael, Smith, 2005: *The Natural Advantage of Nations* (London: Earthscan).
Hawken, Paul; Amory, Lovins; Hunter, Lovins, 1999: *Natural Capitalism* (London: Earthscan).
Heinberg, Richard, 2007: *Peak Everything. Waking up to the century of decline* (Gabriola Island, BC: New Society Publishers).
Millock, Katrin; Cèline, Nauges; Thomas, Sterner, 2004: "Environmental taxes: A comparison of French and Swedish experience from taxes on industrial air pollution", in: CESifo DICE Report, 1/2004.
Mishan, Edward, 1967: *The Costs of Economic Growth* (London: Staples).
Smith, Michael; Hargroves, Charlie, Karlson, 2010: *Cents and Sustainability* (London: Earthscan).
Stern, Nicholas, 2007: *The Stern Review. The Economics of Climate Change* (Cambridge, New York: Cambridge University Press).
World Resources Institute (undated): "Earth Trends Portal: Cumulative CO_2 emissions 19002004", 30 May 2009.

Commencement at the Bren School, June, 2008, with Commencement Speaker Lee Stein (*centre*) and University Chancellor Prof. Henry Yang (*right*), U.C. Santa Barbara. Permission given by James Badham, Bren School

Chapter 18
Climate Sceptics Keep a Distance from Solutions

Climate sceptics abound in the United States.[1] In 2011, 47 % of US citizens preferred to believe that global warming, if it exists at all, results from natural causes.[2] One motive may be a perceived conflict between reducing human greenhouse gas emissions and the American Way of Life. Others, such as Amory Lovins (Lovins/Rocky Mountain Institute 2011), who offer attractive business solutions, have no difficulty agreeing with what 98 % of climate scientists consider to be already proven: global warming is the significant consequence of human influences.

Now sceptics of the opposite sort are raising their voices, saying that all that has been done so far to curb global warming is far too little. Ulrich Hoffmann (2011), admitting that only the EU has been truly active in climate policies, plainly observes that the European Trading System (ETS) has failed to reduce emissions. He shows that European industrial emitters need not take any domestic action until 2017. Generous allocations of permits and incredible offset credits worth 1.6 billion tons of CO_2, mostly under the 'Clean Development Mechanism' (sarcastic name, that!) made this possible. For all its good environmental intentions, the ETS has become a huge subsidy system for polluters. Power companies, according to Hoffmann, made windfall profits from ETS worth 19 billion euros during the first phase, and "are set to rake in up to 71 billion euros in phase II" (Ulrich Hoffmann 2011: 2).

Hoffmann also harbours little hope for solutions. In fact, he is lambasting 'green growth illusions'. To reach what climate experts consider the minimum for maintaining an agreeable climate, a reduction of carbon intensity by a factor of 36 will be needed by 2050, assuming that world population would then peak at nine billion and that current income growth trends will have continued. And if, in

---

[1] This text was initially written for the International Institute for Advanced Sustainability Studies (IIASS) in Potsdam which granted permission to publish this text.

[2] According to a 2011 worldwide Gallup poll (Ray, Julie; Anita Pugliese (2011-04-22): "Worldwide, Blame for Climate Change Falls on Humans"; at: Gallup.com, the US is the only country showing such scepticism.

addition, EU 2007 income levels become standard for all nine billion, Hoffmann calculates that the needed decarbonization factor would rise to 128—a phenomenon entirely unimaginable to any energy expert.

So what could be the solution? In terms of technology, an increase in resource productivity of at least five-fold is attainable over the next three or four decades (Weizsäcke et al. 2009). And that is just the beginning. Of course, efficiency improvements cannot be allowed to be gobbled up by equivalent growth in resource consumption—a phenomenon referred to as the *Jevons Paradox,* or rebound effect. But then, the empirical evidence of the rebound effect stems from times of basically cheap energy and other resources.

This leads us to the political part of the solution. We should stop subsidizing resource consumption. We should see to it that from now on, energy, notably fossil and nuclear, becomes ever more expensive from year to year, and far into the future at that (Weizsäcke et al. 2009, Chap. 9). To avoid hardship and deindustrialization, the strategy should consist of three components:

- Increase prices *parallel to documented efficiency gains*, so that the *average* price paid for energy services remains stable; the success model for this being the self-accelerating and parallel increase of wages and labour productivity over more than 100 years;
- Allow for an affordable *life-line* tariff for the poor; that's agreed practice in South Africa;
- Protect vulnerable industrial branches by recycling the revenues collected from them to the branch, not on a per-gigajoule basis but on a per-job basis, thus *incentivizing job creation and energy efficiency at the same time.*

Even such a moderate price trajectory could effectively stop or starkly reduce the rebound. It would shift wealth creation from resource destruction to services, green industry, and culture. Countries could agree on the strategy by themselves, not waiting for broad international consensus. They would actually benefit from it and gain international competitiveness in a field that plays a rapidly growing role (McKinsey Global Institute 2011). As other countries recognize this dynamic, they will eagerly join, provided they can control their domestic lobbies making money from destroying nature and climate. This will be easier in Egypt than in Saudi Arabia, easier in Germany or the EU than in Canada, easier in China than in Brazil. So let us try to convene an alliance of the winning countries going ahead, becoming richer and thus defining the technological trends for the coming 50 or 100 years. The slower countries will have no other choice but to join later.

References

Amory, B. Lovins; Rocky Mountain Institute, 2011: *Reinventing Fire. Bold Business Solutions for the New Energy Era* (White River Junction, VT: Chelsea Green).

Ernst, von Weizsäcker; Charlie, Hargroves et al., 2009: *Factor Five. Transforming the Global economy through 80 % Improvements in Resource Productivity* (London: Earthscan).

McKinsey Global Institute, 2011: *Resource Revolution. Meeting the world's energy, materials, food and water needs.*

Ulrich, Hoffmann, 2011: *Some reflections on climate change, green growth illusions and development space.* UNCTAD paper #205 (December).

Chapter 19
Decoupling: Technological Opportunities and Policy Options: Executive Summary

19.1 Introduction

As the work of the *International Resource Panel* (IRP) shows, the worldwide use of natural resources has accelerated, bringing with it the thinning or depletion of numerous resource stocks and causing negative environmental impacts (Hertwich et al. 2010). Adjusting our societies to these trends is one of the grand challenges of our times. The trends in resource use suggest that successful economies will be the economies that can increase the value they deliver, while using fewer resources.

This report highlights existing technological possibilities, for developing and developed countries, and the economic advantages. It shows that there is growing evidence that decoupling may be one of the next big opportunities for economic growth, innovation and wise use of resources. The report explores the actions that a country would need to take to create the conditions for its economy to prosper in the future.

It finds that policymakers with leadership, vision and an understanding of political realities can take steps to benefit from future resource trends. The report identifies the barriers that can hold back effective policy change, and examines technological, organisational and policy options that have proved to be successful in different regions of the world. It highlights the forms of policy action that can make faster progress towards the decoupling of economic growth from use of resources.

Co-authored with Karlson 'Charlie' Hargroves et al., for UNEP's International Resource Panel.**Lead Authors**: Ernst Ulrich von Weizsäcker (lead coordinating author),Jacqueline Aloisi de Larderel, Karlson 'Charlie' Hargroves, Christian Hudson, Michael HarrisonSmith, and Maria Amelia Enriquez Rodrigues. **Contributors:** Anna Bella Siriban Manalang,Kevin Urama, Sangwon Suh, Mark Swilling, Janet Salem, Kohmei Halada, Heinz Leuenberger,Cheryl Desha, Angie Reeve, David Sparks.

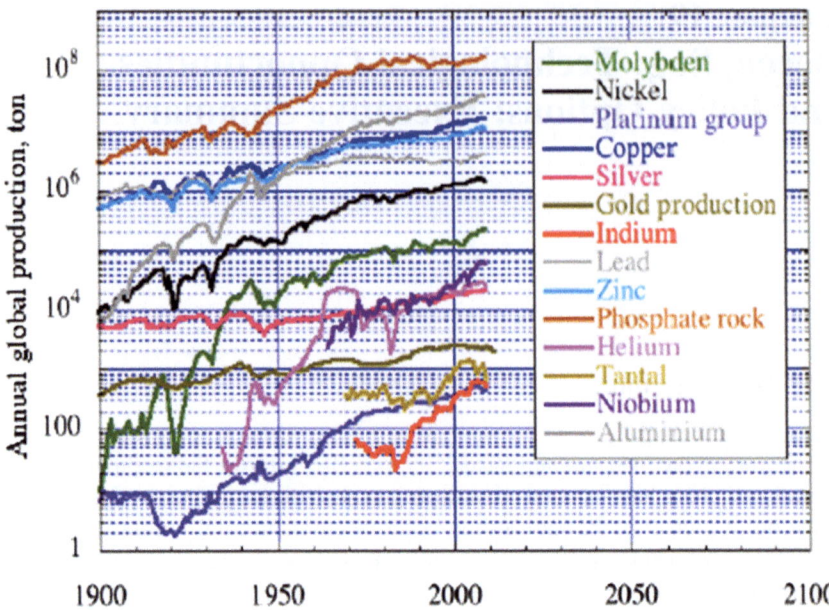

Fig. 19.1 Extraction of many metals grew exponentially since the year 1900 (the ordinate on the picture being logarithmic) From Sverdrup, H.U.; Koca, D.; Ragnarsdóttir, K.V. 2013: "Peak Metals, Minerals, Energy, Wealth, Food and Population: Urgent Policy Considerations for a Sustainable Society", in: *Journal of Enviromental Science and Engineering*, B 2: 189–222

19.2 Trends in Resource Use

During the twentieth century, the annual extraction of ores and minerals grew by a factor of 27, construction materials by a factor of 34, fossil fuels by a factor of 12 and biomass by a factor of 3.6. In total, material extraction increased by a factor of about eight.

The extraction of many metals has followed an essentially exponential growth path since the beginning of the twentieth century, as Fig. 19.1 shows.

Other reports have illustrated that the use of some natural resources essential to prosperity—including freshwater, land and soils, and fish—have similarly increased, in many cases beyond sustainable levels.

The underlying drivers for this explosion in demand appear set to continue. The UN projects global population to grow by more than 2.5 billion people by 2050 (UN DESA 2013) and incomes (on average) are on track to continue rising. According to one estimate, in 20 years there will be 3 billion more people worldwide enjoying 'middle class' income levels, compared to 1.8 billion today (Kharas 2010).

In our first Decoupling report, we described four future scenarios for resource use. In the scenario which represents many policymakers' current plans—in which levels of resource use per head for all global citizens reached the levels of current use of the average European—annual resource extraction would need to triple by

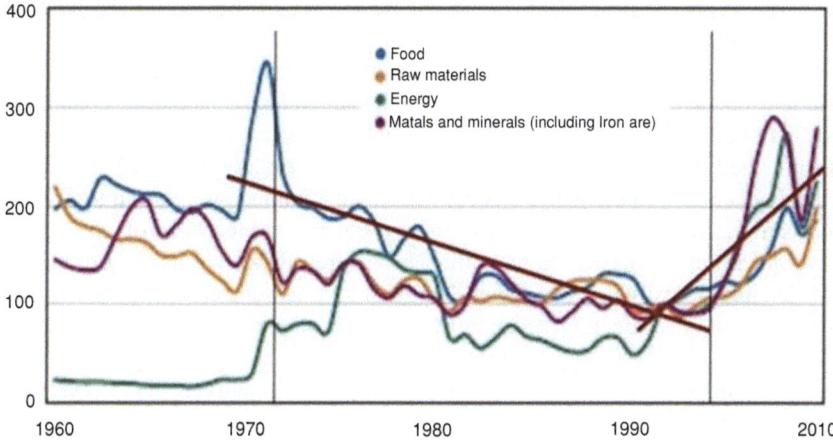

Fig. 19.2 Commodity price indices (*Source* World Bank Commodity Price Data, 2011. Trend lines added, *Courtesy* Swilling, M. World Bank Commodity Price Data (Pink Sheet), historical price data. http://blogs.worldbank.org/prospects/globalcommodity-watch-March-2011

2050, compared to extraction in 2000. This probably exceeds all possible measures of available resources and assessments of the limits of the planet to absorb the impacts of their extraction and use. For example, global demand for water is expected to rise by 40 %, so that in 20 years' time available supplies will probably satisfy only 60 % of world demand (2030 Water Resources Group 2009).

19.3 Consequences of These Changes

It does not seem possible for a global economy based on the current high-consumption model of resources to continue into the future. The economic consequences of increasing resource use are already apparent in three areas: increases in resource prices, increased price volatility and disruption of environmental systems.

Price increases During most of the nineteenth and twentieth centuries, commodity prices had a tendency of declining. But recent developments of massively increased demand have caused the reverse, as shown in Fig. 19.2.

Increased price volatility Price volatility and supply shocks have already been observed across a range of key materials and commodities used in the economy. For instance, the United Nations Food and Agriculture Organization found that the volatility of food prices increased to 22.4 % in 2000–2012 compared to 7.7 % in 1990–1999.[1] Price volatility can be more disruptive than trends of price

---

[1] Measured by the standard deviation around the average price. However, note that before 1990, food prices were also volatile, having a higher standard deviation than in the years 2000–2012.

increase—some believe that rising global food prices led to civil dissatisfaction which fuelled the 'Arab Spring'.[2]

Disruption of environmental systems There are strong links between resource use and damage and depletion of environmental systems, including greenhouse gas emissions (Hertwich et al. 2010). The UN Millennium Ecosystem Assessment documented several accelerating, abrupt, and potentially irreversible changes already occurring to the world's ecosystems, and a number anticipated to occur in the coming decades. These include fishery collapses, bleaching of coral reefs, desertification, increased vulnerability to natural disasters, and crop failures (Millennium Ecosystem Assessment 2005). Studies show that such environmental deterioration is affecting economies and economic growth (Stern 2006; Brown 2008).

There are several reasons why market response is unlikely to respond adequately to these by raising supply of resources to meet demand.

- The scale and rate of change has accelerated, and often outpaces the supply side response.
- There are real physical constraints: past mining of the most attractive ores has led to declining average ore grades for several key metals, such as copper, gold or tin, so that, for many metals, about three times as much material needs to be moved for the same quantity of metal extraction as a century ago.
- For environmental resources—like climate, fish stocks or local ecosystems, too much stress can lead to sudden, non-linear collapse (Smith et al. 2010, Chap. 5, Element 2).
- And, importantly, markets are not adequately set up to factor in much of the expected scarcity of resources—but rather reflect today's extraction cost of still conveniently available ores.

19.4 Strategic Implications

The resource trends have strategic implications for economies. They appear likely to alter the relative importance of resource compared to other inputs into production—and in doing so change the basis of relative competitive advantage between countries. This implies that the economies that move first, or fastest, to adapt to the changed economic conditions stand to gain and bring greater security and wealth to their populations.

To put it in more radical language: as the current model of development is not sustainable in the long term, a real change of course will be needed, changing technologies, policies and consumption habits quite dramatically.

---

[2] For example: *The Arab Spring and Climate Change* (Washington, D.C.: Centre for American Progress, Stimson Centre, The Centre for Climate and Security, 2013).

Some commentators believe that the economic structures of many developing countries mean that they are more able to adapt, and so can gain from change, compared to some more developed countries, that are locked into wasteful infrastructures and habits.

At the same time, trends in resources increase the risks of disruption to economic growth from resource scarcity and shocks, including environmental degradation and collapse. These often cause more severe effects in developing countries, than in richer economies.

19.5 Choices of Response for Policymakers

For economic prosperity and growth, one of the most appealing strategies for adapting is decoupling (Fischer-Kowalski/Swilling 2011; Smith et al. 2010)—the seizing of opportunities for resource productivity, so that a nation can produce greater economic value out of fewer resource inputs (both material and energy) per unit of value.[3] When considering changes, decision-makers need to look as closely as they can at the productivity changes in the resources that matter most to them. Aggregate figures for resource use—which are frequently the most available—may not reflect the possibilities for decoupling economic growth from some particularly important resources.

Decoupling, can mean different achievements. We propose to distinguish between three types of decoupling:

1. *Decoupling through maturation.* This type of decoupling is a 'natural' process of overcoming clumsy techniques, of maturing in the build-up of infrastructures, and of actively reducing environmental pollution.
2. *Decoupling through trade* (burden-shifting). This is related to the maturation process as countries shift from an extraction and production-based economy towards a service economy. If domestic extraction and production is replaced by imported products, resource use will decline domestically, but may still occur elsewhere in the world. This type of decoupling is often labelled as burden-shifting.
3. *Decoupling through intentional resource productivity increase.* This is what is really needed if we want to reduce pressures on limited resources, on climate, and on the environment in general. It requires a lot of technological innovation, infrastructures conducive to resource efficient and low material intensity living and manufacturing, and appropriate attitudes and consumption patterns. Intentional decoupling is the main focus of this report.

Investments in resource productivity can bring multiple gains, ranging from reduced operational costs for companies and the public sector to better environmental quality and the creation of jobs (Smith et al. 2010, notably Chap. 9:

---

[3] Growth is more strongly decoupled where a greater share of an economy's growth comes from resource productivity relative to labour productivity.

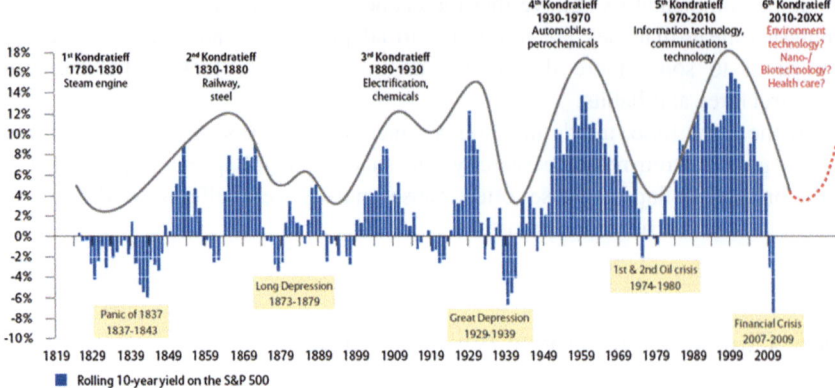

Fig. 19.3 Kondratiev cycles. *Source* Allianz Global Investors 'The Sixth Kondratieff'—Long waves of prosperity, 2010. The description of the sixth Kondratieff suggests that resource productivity could become the overarching characteristic of the new cycle

Decoupling Economic Growth from Freshwater Extraction). For example, energy efficiency policies in California are estimated to have created nearly 1.5 million jobs from 1977 to 2007. Similar figures emerged from Germany's resource productivity policies in the years before 2004, creating or saving more than 1 million jobs.[4]

Economic growth comes, partly, through investments in innovations, and policymakers can influence the nature of the innovations that receive investment through their policies. A vivid visualisation of the relationship between innovation and economic growth is given by 'Kondratiev cycles' (Freeman/Louçã 2001). Economic growth has been observed to come in waves of prosperity, each driven by the spread of new technologies and structural economic change. Figure 19.3 illustrates the way that growth usually involves changes in technologies.

Considering the trends in global resources and environmental resource degradation, we might expect a well-functioning economy to naturally respond to information on resource scarcity by increasing innovations in resource productivity. That implies that decoupling would be one of the drivers of the next period of growth in successful economies.

In practice, there are several barriers and biases that hold back the desired improvements, meaning that the steep rise in resource productivity requires courageous policy changes (Fischer-Kowalski/Swilling 2011: 48, 74). In the past era of declining resource prices, business has tended to focus on increasing labour productivity—with the result that labour productivity has grown at faster rates than other factor productivity (Fig. 19.4).

While the existing policy set may have been suitable for promoting growth in the past, it seems unlikely to meet the challenges of the future. The trends in resources

---

[4] See: Fischer et al. (2004). Also the Ecological Tax Reform 1999–2003 created jobs, chiefly by reducing indirect labour cost.

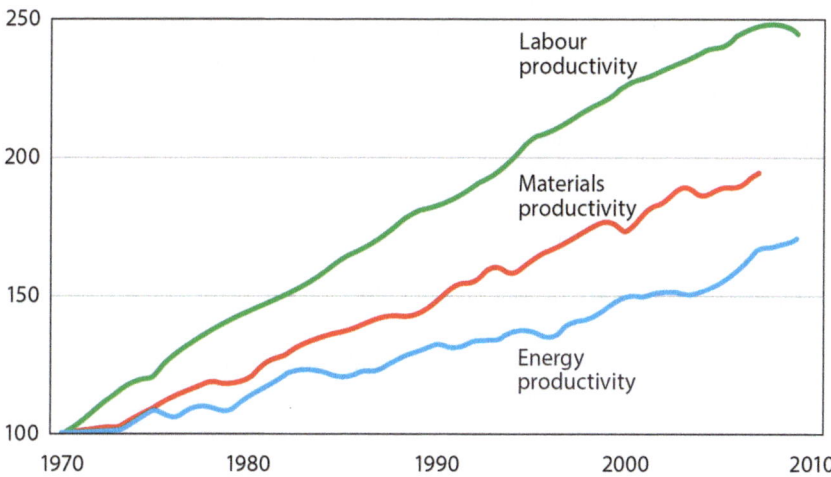

Fig. 19.4 Resource productivity, labour productivity and energy productivity. *Source* EEA 2011

imply that to maintain stable future economies and global, natural life support systems, resource productivity increases would need to be greater than the rate of economic growth for the world as a whole. This is called 'absolute decoupling'.

Indeed, for resources—although pressures differ greatly by resource and country—approximately a factor five improvement (von Weizsäcker et al. 2009) in total resource productivity by 2050 would be required for OECD countries (resulting in just 20 % of today's material usage/unit of production), and the countries that trade with them. This implies that each unit of production is produced using between 25 and 10 % of its current resource inputs by 2050 (World Business Council for Sustainable Development, WBCSD 2010), a much greater rate than resource productivity gains previously seen.

For instance, the Intergovernmental Panel on Climate Change's (IPCC) Fourth Assessment Report, published in 2007, warns that to maintain an agreeable kind of climate, global emissions need to peak by 2015, and then reduce by 25–40 % by 2020 and 80 % by 2050. OECD countries would need to absolutely decouple their growth from their greenhouse gas emissions, at a rate that would give more room to developing countries to raise living standards until they too can achieve absolute decoupling. Apart from greenhouse gas emissions, such decoupling is also needed for a number of other resources such as forestry, fishery, food, waste, air pollution, minerals. The IPCC's Fifth Assessment Report, published in October 2013, also confirms these findings.

The required intentional policy change should influence all aspects of economic and environmental policies, with a view of facilitating their economy's transition to absolute decoupling.

Knowing that relative decoupling will not suffice on a global scale, the focus of this report is on the opportunities for countries to pursue strategies of better lives for their people while significantly reducing resource intensity and, where feasible, even achieving absolute decoupling of resource use. As an encouragement for decoupling policies, our report shows that:

- The potential exists for much greater levels of absolute decoupling to be achieved through strategic changes in technologies and design. Much of the technologies and technique 'know how' to achieve significant levels of resource productivity (as much as five to tenfold improvements) already exists. A number of publications over the last 15 years (Hawken et al. 1999; McDonough/Braungart 2002; Hargroves/Smith 2005; Pacala/Socolow 2004; Pauli 2010; Smith et al. 2010; Lovins 2011) have shown that decoupling is technically possible for material resource consumption, greenhouse gases, and water extraction (Chap. 3).
- Success stories exist of countries that achieved some modest absolute decoupling of economic growth from selected aspects of resource use and greenhouse gas emissions, from which we can learn (Chaps. 6 and 7).
- Much of the policy 'know how' required to achieve economy wide 'decoupling' exists in the form of legislation, incentive systems, administrative measures, and institutional reform. But additional policy options could be opened for a yet more strategic and long-term avenue towards ecologically sound growth (Chaps. 7 and 8).

19.6 Technological Responses Allowing Significant Decoupling

Increasing resource productivity is technologically possible: technologies and techniques that bring very significant resource productivity gains are already available, right across the range of resource consuming activities, with different technologies applicable at different levels of economic development.

The Rathkerewwa Desiccated Coconut Industry (RDCI) in Maspotha, Sri Lanka, provides a good example. RDCI could reduce 12 % of energy use, 8 % of material use and 68 % of water use, while increasing the production by 8 % during the same period by adopting a series of recommendations on its peeling process, water treatment, and fuel switching. The total investment required for implementing these recommendations was less than US$5,000, while an annual financial return of about US$300,000 was reported.[5] Sweden introduced an energy efficiency programme in 2005 for its energy intensive industries. A recent analysis showed average payback periods of less than 1.5 years (Stenqvist/Nilsson 2013).

---

[5] For details, see at: http://www.unido.org/fileadmin/user_media/Services/Environmental_Management/Cleaner_Production/RECP_SriLanka.pdf.

19.6 Technological Responses Allowing Significant Decoupling

The scale of the opportunity is very large. One estimate places the savings potential between US$2.9 trillion and US$3.7 trillion each year (by 2030). 90 % of the opportunities had an internal rate of return of greater than 10 %, if adjusted for subsidies, carbon prices and a social discount rate.[6] The following examples provide an illustration of some of the potential:

High-efficiency motors These could potentially save 28–50 % of motor energy use, with a typical payback period of one to three years (CADDET 1995). Electric motors used in industry in China account for around 60 % of the country's total electricity consumption. The operational efficiency of these motors is 10–30 % below international best practice, depending on the industry. A pilot study at China's second-largest oil field suggested there was the potential to save more than 400 million kilowatt hours (kW h) of electricity per year in the oil field, with a payback period for recovery of the initial investment of 1.6 years (UNEP 2010).

Higher strength steel Using steel with higher strength for re-enforcement of concrete, beams and columns saves steel: ArcelorMittal, the world's largest steel company estimates use of higher strength steel achieves a 32 % reduction in the weight of steel columns and 19 % in beams (McKinsey Global Institute 2011: 105). China and developing countries tend to use lower-strength steel, with China using steel for reinforcement that is two-thirds the strength of steel averagely used in Europe. This offers a significant opportunity—as these countries' use of steel is very significant. (For example, China currently consumes 60 % of global steel reinforcement bar production.) Even partial global switching to higher strength steel could save 105 million tonnes of steel a year, and save 20 % of the costs of the use of steel.[7]

Blanking sheet metal The pressing out (or 'blanking') of metal components of different size and shape from sheet metal necessarily leaves behind pieces of sheet metal that are not wanted and too small to use for other components. Intelligent organisation of the different shapes to be pressed out can realise significant metal savings. Deutsche Mechatronics GmbH operates in Germany using computer-driven shuffling and a good production planning system that could reduce metal use by 12 %.

Methane from waste landfill In the United States of America (USA), approximately 480 landfill sites, representing around 27 % of the nation's landfills, capture released methane gas from decomposing organic waste (2009 figures) (Bracmort et al. 2009). It is estimated that between 60 and 90 % of the methane in the landfill gas can be captured and burnt. Nevertheless, methane from landfills contributes 1.8 % to the US total greenhouse gas emissions.

Drip irrigation Agriculture is responsible for 70 % of freshwater withdrawals (von Weizsäcker et al. 2009). In many countries, 90 % of irrigated land receives irrigation water through open channels or by intentional flooding. The waste of freshwater through these methods, through evaporation, leakage and seepage is high. Farmers in

---

[6] McKinsey Global Institute (2011). The study suggests also that 70 % of the opportunities have a greater than 10 % IRR at current prices. The higher figure (3.7 trillion) applies if carbon is appropriately priced and perverse subsidies phased out.
[7] *Sustainable Materials: With Both Eyes Open*, l.c.: 178.

India, Israel, Jordan, Spain and the USA have shown that sub-surface drip irrigation systems that deliver water directly to crop roots can reduce water use by 30–70 % and raise crop yields by 20–90 %, depending on the crop (Postel et al. 2001). Efficiency savings can be as high as 50–80 %, and can be made affordable for use in the developing world (Shah/Keller 2002) with payback periods of less than a year.

19.7 Creating the Conditions for Investments in Resource Productivity

Policymakers can facilitate the widespread uptake of technologies and techniques for decoupling. A wealth of experience from past policies on innovation, decoupling and environment can guide future policy action. Lessons can be learned from some great successes: for example in water efficiency. In Australia, GDP rose by 30 % and water consumption was reduced in absolute terms by 40 % from 2001 to 2009 (Smith et al. 2010, l.c.). Singapore's spectacular development brought 25-fold growth over 40 years while water use grew only fivefold. This implies a fivefold increase of water productivity over that period (Khoo 2005).

Many countries have put in places policy mixes promoting decoupling. For example, at European Union (EU) level, recent initiatives, such as the 7th Environmental Action Programme and the Roadmap to a Resource Efficient Europe, and the Energy Efficiency Directive of 2012 are long-term strategies moving energy, climate change, research and innovation, industry, transport, agriculture, fisheries and environment policy all towards decoupling. The roadmap also deals with tax policy, making the case for a shift from labour taxes to resource taxes, and discusses the phasing out of environmentally harmful subsidies. Similarly, China has strategically improved energy efficiency writing 20 and 16 % efficiency gains into its eleventh and twelfth 5-Year Plans respectively, and adopting regulation and incentives to make it happen.

Whether, where and how decoupling occurs may depend on national decision makers' abilities to overcome biases which currently disadvantage investments resource productivity. Countries that can overcome those barriers can lead the next wave of transition, and gain advantage over their competitors.

19.8 Changing Current Biases

There are currently several factors that lead to bias against investments in resource productivity and two areas of barriers for policymakers to tackle. The first group arises from the effect of the historic policy framework. There a number of areas where current policy structures coming out of past government decisions steer economies away from resource productivity, examples of which are:

- Subsidies of up to US$1.1 trillion each year for resource consumption (McKinsey Global Institute 2011, ibid). These subsidies encourage the wasteful use of resources while reducing the savings from investments to use the resources more efficiently.
- Taxation of people's work through labour taxes is typically higher than the tax burden on resources (and energy). As labour and resources are often alternative inputs into economic growth, this favours resource consumption rather than increased employment. Together with distortions from subsidisation of resources, taxation reduces the return on investment in resource efficient technologies and techniques. Taking the economy as a whole, it encourages development of an economy that is more resource intensive than it needs to be.
- Regulatory frameworks for markets have often been created in ways that discourage long-term management of resources, but rather promote their wasteful early use. Market regulations that have worked well for old technologies may disadvantage the entry of new technologies. For instance, in some developed country energy markets, bidding systems for electricity supply have taken place one day in advance of electricity delivery. This has put operators of wind turbines at a disadvantage, because they can only reliably predict their electricity output three hours in advance (OECD 2010).

The second group of factors holding back decoupling are biases against change. These can be seen as physical and technological biases, behavioural biases, organisational and institutional biases.

- Technological bias can arise because many technologies are used in conjunction with existing physical infrastructure, giving existing technologies a significant advantage over alternative technologies that would require different infrastructure (for example, the lack of electric vehicles' recharging points compared to the large number of refuelling stations for oil-powered vehicles).
- Organisational and institutional biases arise from the way in which standard practices, cultural norms, accepted wisdom and rules influence peoples' behaviours and the decisions they make. To illustrate this with one example from the finance sector: due to the internal incentives and controls found in many banks and financing organisations, positive financing decisions tend to be made in areas familiar to the professional expertise of staff. The lack of track record for the investment performance of new technologies makes them appear more risky, and places them at a severe disadvantage when investment decisions are made (Hudson et al. 2013). This represents a problem as meeting the world's future consumption demands through resource efficient technologies (or supply side technologies) has been estimated to require around US$3 trillion of investment a year globally (McKinsey Global Institute 2011, ibid) for which the financing will need to be found.

Both these groups of barriers need to be tackled to make full progress to a successful, resource-productive society. Policy changes can overcome these barriers. In doing so, it would create conditions where investments in resource productivity became more attractive than alternative investments, and open up the universe of opportunities offered by decoupling.

19.9 'Lock-In' to Political and Economic Structures

Relatively few opportunities for beneficial policy change are currently taken up. Part of the reason for this seems to be that political systems have their own inertia, which often act as a brake on policy reform, or block it entirely. The close interaction in nearly all countries between political decision-making and economic interests can lead to what is called 'systems lock-in' because the policy framework is difficult to change without change to economic interests and vice versa. Political processes can therefore act as barriers to decoupling, because:

- Frequently, policy is formed in response to the interests of leading economic groupings. Where these groupings are biased towards the current arrangements that have given them market power, they tend to engage strongly to preserve existing policy. This can be the case even as underlying conditions change (like resource availability).
- Segmented policy-making governmental structures—with different ministers or departments favouring different specific interest groups—lead to policy inconsistency, with the effect of some policies being cancelled out by the indirect effect of others. This inconsistency, lack of clear direction and past records of changes in policy creates unpredictability and uncertainty about future investment return dependent on lasting policy change.
- The institutions through which policies are made often reflect existing norms, and change is often resisted, within the institutions (for example government departments) or industrial organisations shaping policy (Ekins/Salmons 2010, Chap. 5: 132).
- Where economic interests are at stake, groups are likely to contest evidence showing the need for change. Where there is some degree of scientific uncertainty about the future (as is inevitable) this can be used to discredit unfavourable information. Even evidence gathered by governments seeking to promote innovation may be sceptically received and scrutinised for bias. This rejection of, or unwillingness to hear, information demonstrating the benefits of change is a key barrier to achieving policy change—as success in policy reform often involves political and economic actors perception of their own self-interest to alter (Ekins/Salmons 2010, ibid: 133–134).
- Policymaking procedures are often lengthy, and can have additional lead in-times before policy is expected to take effect—leading to lags in the policy framework in reaction to new information.

The inertia created by these political and procedural factors is frequently the primary barrier to successful decoupling. Understanding these aspects of the problem can assist policy makers in making further progress.

19.10 Making Progress with Resource Productivity

Policy change, in the face of this significant inertia, requires leadership. A central part of this leadership will be a clear vision of a successful future economy, well adjusted to trends in resource use and scarcity. Many different policy changes can create these favourable conditions—Chap. 7 of this report gives some illustrations of past and current policies. So, there are opportunities for leadership for many people. This includes individuals working within organisations and institutions across most parts of government, the economy and civil society (including consumers). Inside government, there are opportunities for decision-makers with influence on policies regarding industry, development, innovation, environment, employment, and taxation.

This wealth of options for areas for positive change arises because decoupling is often best stimulated by creating favourable conditions for investment in resource productive innovation, and letting market forces provide the best solutions. For these kind of changes, there is clearly no 'one size fits all' prescription or instrument, but some common features can be identified for policies aiming at ambitious goals of decoupling.

19.11 Unlocking Change in Policies

Replacement, reform or complementary addition to parts of the old policy framework, and the reduction of the biases against decoupling is possible, and has been often been achieved. Success in creating the conditions for decoupling would need to unlock the observed resistance to policy reform. In this task, the chances of success appear higher where the policymaker looks at the institutional framework in which the political decision is made. In practice for changes to policy, this means being aware of the set of actors who are able to influence the decision, their interests, relative power and the norms and assumptions which are shaping the decision. Those seeking change:

> … need to become adept at institutional analysis, identifying those elements supportive, or hostile to, the reform in question, and work to strengthen the more supportive elements and weaken the more hostile ones (Ekins/Salmons 2010, Chap. 5: 132).

For example, there are frequently synergies between policies for decoupling and other policy goals. These can be used to win support for policy change. This was the case in Germany which introduced a relevant tax reform from 1999–2003 in five consecutive steps, eventually shifting some €18 billion annually from indirect labour charges to taxes on energy. One motive for the tax reform was to reduce incentives for environmental harm, but it also allowed the corresponding reduction of other taxes on labour that lead to an estimated gain of 250,000 jobs (Knigge/Görlach 2005). The World Bank's summary of benefits from an

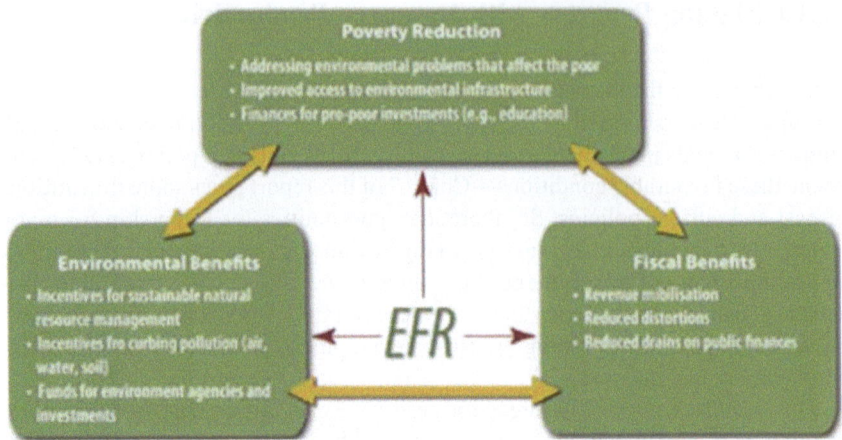

Fig. 19.5 Assumed benefits from an Environmental Fiscal Reform (EFR). *Source* World Bank 2005, l.c.: 18

environmental fiscal reform (World Bank 2005) gives one illustration of the potential achievement of multiple goals (Fig. 19.5).

Based on past experience with policy changes (Ekins/Salmons 2010, Chap. 5), success in decoupling appears to be more likely where policymakers seeking change:

- Take account of the potential losers from policy change, and consider what will bring enough of them to favour change.
- Help those affected by change to focus their innovation towards a consensus future goal, by changing their expectations of the future. By creating shared visions and credible strategies, future investment patterns can be changed, often without great expense, as firms shift in advance to profit from new conditions.
- Create, or rely on, a source of sufficiently trusted independent advice—on the science or on the impacts of change. Objective, transparent scientific evidence is very useful: information sources seen to be self-interested will be much less effective.
- Present concrete examples of policies or practices used in different countries, or in different realms of policy. Many of the reforms to increase decoupling will require new structures, behaviours or business models that may seem initially unfamiliar, and odd. Demonstrating that different arrangements work elsewhere can be convincing.
- Create an institutional structure for the specific policy decision that is participatory, sufficiently broad to contain enough people who can form a pro-reform coalition and set up in a way that allows potential supporters of change to voice their support. This facilitates information flows, and can help form a common vision for the future that reconciles previously opposing views.

- Use a simultaneous mix of policy instruments. This can help the actors in a value-chain of economic activity (for example, from raw material extraction to final product consumption and recycling) to change profitably together. This may be necessary to overcome a 'lock-in' between demand and supply, which can commonly happen when a seller offers what is being demanded, the purchaser buys what is being offered and there is little scope for either to innovate.
- Work to increase the cumulative effect of several smaller steps, as it is rarely the case that political or economic conditions exists that allow a policymaker to bring about a very large, radical change in resource productivity in one step.
- Be aware of options for reform and use political opportunities when they arise. Good economic times are often more favourable for introducing change, with less fear of negative consequences and greater availability of finance for innovative investments. Yet, crises can also facilitate reform, in different ways. ...

19.12 Changing the Institutional Framework to Facilitate Future Policy Reform

One aspect of successful reform is to take steps that create the conditions for further, future policy reform. Making changes to decision-making processes, either internal to an organisation or external, can indirectly facilitate future change.

In government, this could mean making a change to the decision-making structures (like the mandate of ministers or committees) that allows decisions promoting the long-term management of resources to be taken more easily. It could also mean implementing a policy that increases the future economic and political weight of innovators, or favourably changes the perception of potential opponents to change (for example by changing company reporting to include information on resources that helps companies take resource factors into account in their business decisions).

Changes to institutional decision-making structures have long been appreciated to have important beneficial outcomes, and this is particularly the case for overcoming the bias of decision-making towards the short term.

For example, the UK is seen as a strong, liberal economy. In part this is because, in 1998, authority over monetary policy was passed from the government to the central Bank of England. This transferred the power to set interest rates—a power of huge importance to the economy. The aim was to provide greater economic stability by distancing those decisions from short-term political influence.

There have also been many examples where international agreements have acted as stimulation for domestic action. In part this is because concerted action between countries, which reduces fears of unfavourable distortions in international markets. But it is also because an international commitment can act as a persuasive tool against opponents of change, not least by indicating that change is viewed as internationally important.

19.13 Putting Decoupling into Practice: Linking Resource Price Rises to Resource Productivity Gains

Economic instruments to push technologies and markets towards higher resource productivity typically run into one characteristic difficulty: if price signals are strong, industries may just give up or emigrate, and consumers tend to fight the government imposing painful price signals. But if price signals are weak, there is a high likelihood of effects remaining insignificant.

A potential way out is a price signal that steadily increases at the pace of decoupling successes. For example, if the average efficiency of the car fleet rises by 1 % in one year, a 1 % price increase of petrol at the pump would seem fair and tolerable. However, the firm announcement of the continuation of this scheme will induce car manufacturers and traders as well as consumers to speed up efforts to reduce petrol consumption per kilometre or to avoid unnecessary trips. Hence a small signal can have a strong impact if continued over a long period of time.

A policy of this kind can combine several of the considerations to unlock inertia described above, and may come close to the type of combined policy which is needed.

One proposal for a policy could use taxation or subsidy reduction to move the price of a chosen resource upwards in line with documented increases of energy or resource productivity. In the sections below we look at different qualities of this proposal. In practical terms, one would not prescribe an exact price trajectory but a 'corridor' within which prices can fluctuate a little. Interventions would only be made when such fluctuations are leaving the corridor. Interventions can also reduce prices or taxes if fluctuations leave the corridor upwards. The main purpose is predictability so that investors, manufacturers, and consumers know what is going to happen.

19.14 Broadening the Economic Discourse

By establishing a 'ping-pong' between price rise and efficiency gains, costs (which are what influences competitiveness and livelihoods) would, on average, not increase. Under the 'ping-pong' policy, on average, one would pay the same amount of money for the same quality of energy services as during the year before—paying a higher price for each unit of energy, but consuming fewer units of energy, as each unit of energy delivers more output thanks to the productivity gain. Of course, some industries and some families cannot increase their resource productivity as fast as the average gains take place. Politics will have to address this problem by a balanced mix of support measures or exemptions without destroying the incentive to innovate or adapt.

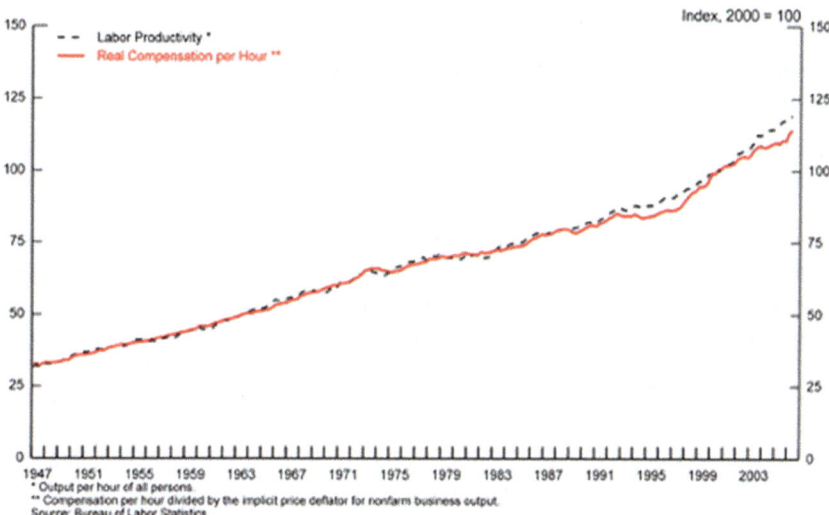

Fig. 19.6 The parallel increase of labour productivity and of gross hourly wages in the United States of America from 1947 to 2007. *Source* US Bureau of Labor

19.15 Creating a Vision of the Future and Reducing Uncertainty

The proposal would not entirely remove uncertainty about returns on investments in resource productivity, as variations in resource prices and uncertainty about future energy or resource productivity increases would remain. However, uncertainty would be reduced, in particular long-term uncertainty about the direction of prices. This would serve as a strong and predictable incentive to investors, states, individual companies or research laboratories to systematically invest in ever more resource productivity. It seems plausible that the mutual reinforcement between prices and efficiency increases will lead to a long term and ultimately dramatic increase of resource productivity.

An interesting partial analogy exists to the proposed 'ping-pong' dynamics between resource productivity gains and resource prices. It is the increase over at least 150 years in labour productivity and gross wages per hour of work. As productivity increased, workers could successfully demand higher wages. And as wages went up, employers were driven to speed up further increases of labour productivity. Figure 19.6 shows the parallel dynamics between labour productivity and wages in the USA over 60 years.

Obviously, the analogy is far from perfect. Wage negotiations typically occur without any state intervention, while the increase or moderation of energy prices does require such interventions. And it is not clear to what extent higher resource prices might lead to moving operations to other countries; in the case of rising wages this is less likely to occur because other countries tend to show the same dynamics of wages rising with productivity.

19.16 Creating Sufficient Winners in Favour of Change

The proposal has aspects that give it the potential to create sufficient winners to form a coalition that supports its introduction. It would provide a source of government revenue, creating choices for the government to reduce taxation on other people or firms in the economy, increase spending or to reduce fiscal deficits. Linking the size of the tax to productivity increases means that the total potential revenue does not decline, even as the number of units of resource consumed decreases.

Secondly, by increasing resource tax at the rate of average efficiency gain, the proposal increases the relative competitive advantage of firms which have above average resource productivity gains: these firms reduce costs relative to their competitors. This not only provides greater incentives for competition based on increased resource productivity, but also provides reasons for the more innovative and productive firms to take political positions in favour of change.

19.17 Taking Account of Potential Losers in a Policy Mix

Introducing a slow, incremental, long-term increase of prices in the way suggested might allow industry and families to gradually adapt to higher price levels and yet would serve as a strong signal for all long-term investments and decisions. Often the signalling effect alone induces more resource-efficient behaviours, as firms and people adjust in anticipation.

The generation of revenues allows some recycling of those revenues to the losers from the policy change. Following a model from Sweden's tax on nitrous oxides, the revenues from the policy could be returned to clusters of firms (such as the non-ferrous metals industries)—not per energy unit consumed but per job added.

Countries have also found ways to protect vulnerable low-income people (who have limited capacity to improve their resource use) from policy-induced price rises. In many countries of the world, a move from generally low and subsidized energy and water prices to realistic market prices (encouraging private capital to invest in more supplies) has been accompanied by policies that allow for a preferential low price level for poor families. South Africa has set a good example within its integrated water plan.

19.18 Creating New Institutional Arrangements

The design of a policy mechanism that raised prices of energy or resources in line with efficiency increases would require new, presumably legally binding, institutional arrangements. Those would be context specific to autonomous countries, but would be likely to involve binding pre-commitment of government to the mechanism, with independent and credible mechanisms for monitoring and calculating documented efficiency gains.

References

2030 Water Resources Group, 2009: *Charting our Water Future: Economic Frameworks to Inform Decision Making* (Munich: 2030 Water Resources Group): 7.
Bracmort, K.; et al., 2009: *Methane Capture: Options for Greenhouse Gas Emission Reduction* (Washington D.C.: Congressional Research Service).
Brown, L.R., 2008: *Plan B 3.0: Mobilizing to Save Civilization* (New York: W.W. Norton & Company)
CADDET, 1995: *Saving Energy with Electric Motor and Drive* (Sittard: CADDET Energy Efficiency).
Ekins, P.; Salmons, R., 2010: *Making Reform Happen: Lessons from OECD Countries* (Paris: OECD).
Fischer, H.; Lichtblau, K.; Meyer, B.; Scheelhaase, J. 2004: "Wachstums- und Beschäftigungsimpui",in: *Wirtschaftsdienst*, 84,4: 247–254.
Fischer-Kowalski, M.; Swilling, M., 2011: *Decoupling Natural Resource Use and Environmental Impacts from Economic Growth* (Nairobi: UNEP)
Freeman, C.; Louçã, F., 2001: *As Time Goes By. From the Industrial Revolution to the Information Revolution* (Oxford: Oxford University Press).
Hargroves, K.; Smith, M., 2005: *The Natural Advantage of Nations: Business Opportunities, Innovation and Governance in the 21st Century* (London: Earthscan—James & James).
Hawken, P.; Lovins, A.; Lovins, L.H., 1999: *Natural Capitalism: Creating the Next Industrial Revolution* (London: Earthscan).
Hertwich, E.; van der Voet, E.; Suh, S.; Tukker, A.; Huijbregts, M.; Kazmierczyk, P.; Lenzen, M.; McNeely, J.; Moriguchi, Y., 2010: *Assessing the Environmental Impacts of Consumption and Production. Priority Products and Materials* (Nairobi: UNEP).
Hudson, C.; Shopp, A.; Neuhoff, K. 2013: *Financing of Energy Efficiency: Influences on European Public Banks' Actions and Ways Forward* (Berlin: Deutsche Institut fur Wirtschaftsforschung).
Kharas, Homi, 2010: *The Emerging Middle Class in Developing Countries*, OECD Development Centre Working Paper 285 (Paris: OECD).
Khoo, T.C., 2005: "Water Resources Management in Singapore", Paper for the 2nd Asian Water Forum, Bali, Indonesia, 29 August–3 September 2005
Knigge, M.; Görlach, B., 2005: *Effects of Germany's Ecological Tax Reforms on the Environment, Employment and Technological Innovation*, Summary of the Final Report of the Project: Quantifizierung der Effekte der Ökologischen Steuerreform auf Umwelt, Beschäftigung und Innovation, Research Project commissioned by the German Federal Environmental Agency (UBA).
Lovins, A., 2011: *Reinventing Fire* (Snowmass: Rocky Mountain Institute).
McDonough, W.; Braungart, M., 2002: *Cradle to Cradle: Remaking the Way We Make Things* (San Francisco: North Point Press).
McKinsey Global Institute, (2011) *Resource Revolution: Meeting the world's energy, material food and water needs*.

Millennium Ecosystem Assessment, 2005: *Ecosystems and Human Well-being: Synthesis* (Washington D.C.: Island Press).

OECD, 2010: *Smart Grids and Renewable Energy, Competition Committee Roundtable* (Paris: OECD).

Pacala, S.; Socolow, R., 2004: "Stabilization Wedges: Solving the Climate Problem for the Next 50 years With Current Technology", in: Science, 305(13 August): 968.

Pauli, G., 2010: *The Blue Economy. 10 Years, 100 Innovations 100 Million Jobs* (AOS: Paradigm Publishers).

Postel, S.; Polak, P.; Gonzales F.; Keller, J., 2001: "Drip Irrigation for Small Farmers: A New Initiative to Alleviate Hunger and Poverty" in: Water International, 26,1: 8

Shah. T.; Keller, J., 2002: "Micro-irrigation and the Poor: Livelihood Potential of Low-Cost Drip and Sprinkler Irrigation in India and Nepal", in: FAO (Ed.): *Private Irrigation in Sub-Saharan Africa* (Rome: FAO—International Water Management Institute): 165.

Smith, M.; Hargroves, K.; Desha, C., 2010: *Cents and Sustainability: Securing Our Common Future by Decoupling Economic Growth from Environmental Pressures, The Natural Edge Project* (Earthscan, London).

Stenqvist, C.; Nilsson, L.J., 2013: "Energy Efficiency in Energy-Intensive Industries—An Evaluation of the Swedish Voluntary Agreement PFE", in Energy Efficiency, 5,2(May): 225–241.

Stern, N., 2006: *The Stern Review* (Cambridge: Cambridge University Press).

UN DESA, 2013: *World Population Prospects: The 2012 Revision* (New York: UN)

UNEP, 2010: *Training Programme on Energy Efficient Technologies for Climate Change Mitigation in Southeast Asia, Case Studies on Electric Motors* (Bangkok, Thailand: United Nations Environment Program).

von Weizsäcker, E.; Hargroves, K.; Smith, M.; Desha, C.; Stasinopoulos, P. 2009: *Factor 5: Transforming the Global Economy through 80 % Improvement in Resource Productivity* (London: Earthscan).

World Bank, 2005: *Environmental Fiscal Reform: What Should be Done and How to Achieve It* (Washington D.C.: World Bank).

World Business Council for Sustainable Development, WBCSD, 2010: *Vision 2050* (Conches-Geneva: WBCSD).

UNEP's International Resource Panel

The International Resource Panel was established in 2007 to provide independent, coherent and authoritative scientific assessment on the sustainable use of natural resources and the environmental impacts of resource use over the full life cycle. By providing up-to-date information and best science available, the International Resource Panel contributes to a better understanding of how to decouple human development and economic growth from environmental degradation. The information contained in the International Resource Panel's reports is intended to be policy relevant and support policy framing, policy and programme planning, and enable evaluation and monitoring of policy effectiveness.

Members and Partners of the International Resource Panel
Co Chairs

- Ernst Ulrich von Weizsäcker, former Chairman, Environment Committee, Bundestag
- Ashok Khosla, former President of IUCN, founder, Development Alternatives, India

Steering Committee

The Steering Committee, consisting of representatives from governments, the EC, UNEP, OECD and other organisations, advises on annual work programmes and budgets.

Members of the Steering Committee as of January 2009 are as follows:
- UNEP Division of Technology, Industry and Economics (co-chair)
- DG Environment, European Commission (co-chair)

Civil Society Organizations

- International Council for Science (ICSU)
- International Chamber of Commerce (ICC)
- International Union for Conservation of Nature (IUCN)
- World Business Council for Sustainable Development (WBCSD)

Secretariat

The Panel is supported by a Secretariat, hosted by the Sustainable Consumption and Production Branch of UNEP's Division of Technology, Industry and Economics, in Paris, France.

Resource Panel Secretariat

United Nations Environment Programme
   Division of Technology, Industry and Economics
   Sustainable Consumption and Production Branch
   15 rue de Milan
   75441 Paris Cedex 09, France
   Email resourcepanel@unep.org
   Website http://www.unep.org/resourcepanel/Home/tabid/106603/Default.aspx

Club of Rome

The Club of Rome was founded in 1968 as an informal association of independent leading personalities from politics, business and science, men and women who are long-term thinkers interested in contributing in a systemic interdisciplinary and holistic manner to a better world. The Club of Rome members share a common concern for the future of humanity and the planet.

The Club of Rome is a non-profit organisation, independent of any political, ideological or religious interests. Its essential mission is "to act as a global catalyst for change through the identification and analysis of the crucial problems facing humanity and the communication of such problems to the most important public and private decision makers as well as to the general public." Its activities should: "adopt a global perspective with awareness of the increasing interdependence of nations. They should, through holistic thinking, achieve a deeper understanding of the complexity of contemporary problems and adopt a trans-disciplinary and long-term perspective focusing on the choices and policies determining the destiny of future generations." The aims of the Club of Rome are: to identify the most crucial problems which will determine the future of humanity through integrated and forward-looking analysis; to evaluate alternative scenarios for the future and to assess risks, choices and opportunities; to develop and propose practical solutions to the challenges identified; to communicate the new insights and knowledge derived from this analysis to decision-makers in the public and private sectors and also to the general public and to stimulate public debate and effective action to improve the prospects for the future.

The Club of Rome, in its early years, focused on the nature of the global problems, the 'problematique', on the "limits to growth" and on new pathways for world development. The Club of Rome is focusing in its new programme on the root causes of the systemic crisis by defining and communicating the need for, the vision and the elements of a new economy, which produces real wealth and wellbeing; which does not degrade our natural resources and provides meaningful jobs and sufficient income for all people. The new programme will also address underlying values, beliefs and paradigms.

The Club of Rome publishes its findings in reports and, since its founding, has released 33 reports concerning the future of humanity. The Club of Rome currently consists of approximately 100 individual members; over 30 national and regional associations; the International Centre in Winterthur, a European Support Centre in Vienna and the Club of Rome Foundation, which provides the opportunity for major individual donors to be involved, to participate in the development and dissemination of the Club's projects and messages.

The activities of the Club are guided by the General Assembly of its members which meets once a year. At present the Club has two Co-Presidents, Anders Wijkman (Sweden) and Ernst Ulrich von Weizsäcker (Germany), and one Vice-President, Dr. Roberto Peccei. The work of the International Club is supported by a small International Centre in Winterthur, Canton Zurich, Switzerland under the leadership of Ian Johnson (United Kingdom).

Address Club of Rome, International Centre, Lagerhausstrasse 9, CH-8400 Winterthur (Can-ton Zurich), Switzerland.
Email info@clubofrome.org
Website http://www.clubofrome.org/

Federation of German Scientists

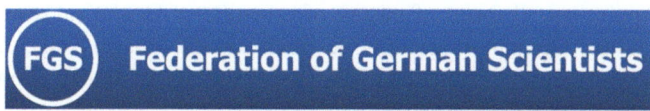

The Federation of German Scientists (FGS; German acronym VDW) was founded in 1959 in West-Berlin by renowned nuclear scientists, including Carl Friedrich von Weizsäcker and the Nobel Prize laureates Max Born, Otto Hahn, Werner Heisenberg, and Max von Laue.

Founding members: G. Burkhardt, C.F. v. Weizsäcker. W. Gerlach

Two years earlier this group of experts had become well-known to the public as "Göttinger 18": Nuclear scientists who had publicly argued against a nuclear armament of the German Bundeswehr. Since then the FGS feels bound to the tradition of responsible science. It has nearly 400 members from different fields of the natural sciences, the humanities, and social sciences, so that a large range of topics is approached at a high level of competence. With the results of its interdisciplinary work the Federation of German Scientists not only addresses the general public, but also the decision-makers at all levels of politics and society.

The members of FGS stand in this tradition. They feel committed to taking into consideration the possible military, political, economic and social implications and possibilities of atomic misuse when carrying out their scientific research and teaching.

In Annual Conferences and in interdisciplinary Expert Groups as well as public comments it addresses issues of science and technology on the one hand, and peace and security policy on the other. At the same time, the role of science itself in genesis and in solution of socio-technological problems is subject of examination and expertise. FGS' membership lists also include representatives of the humanities and social sciences, so that a large range of topics is approached at a high level of competence. With the results of its interdisciplinary work the Federation of German Scientists not only addresses the general public, but also the decision-makers at all levels of politics and society. According to its statutes of 1959, the FGS aims to

- keep up and deepen the awareness of those working in science for their responsibility for the effects which their work has on society;
- study the problems which result from the continuous development of science and technology;
- assist science and its representatives in making public the questions related to the application of scientific and technical developments;
- provide advice and thus exercise influence on decisions as long as they are assessable and can be dealt with by means of scientific knowledge and methods, and to point out all forms of misuse of scientific and technical results;
- to defend the freedom of scientific research and the free exchange of its results and to expand and strengthen the traditional international cooperation of scientists.

Throughout his career, Prof. Dr. Ernst Ulrich von Weizsäcker has been closely affiliated with the VDW, as a member of its board (1985–1991, 2003–2009), as its chairman (1988–1991) and as a member of the VDW's Advisory Committee (2009–2015).

Address Vereinigung Deutscher Wissenschaftler (VDW), Marienstr. 19/20, 10117 Berlin, Germany;
Email info@vdw-ev.de;
Website http://www.vdw-ev.de/index.php/de-DE/

Wuppertal Institute on Environment, Climate and Energy

The Wuppertal Institute was founded in 1991 by Johannes Rau, the then prime minister of the Land North Rhine-Westphalia. With the end of the Cold War and German reunification decision makers became aware of global climate change caused by humankind as a new global challenge. The German Bundestag had established the Parliamentary Study Commission "Protecting Earth's Atmosphere" in 1987, and the need for studying this issue in depth had become apparent. The energy industry is a significant part of the economy of North Rhine-Westphalia; so efforts to reduce greenhouse gas emissions have great impacts. How did industry, the transport sector and private households use energy? What were the alternatives to coal-based electricity generation? These questions would have to be addressed if solutions acceptable to broad sectors of society and involving both the supply and demand sides were to be found. The Wuppertal Institute was established to do just that. The founding president, Professor *Ernst Ulrich von Weizsäcker* who directed it until 2000 started with a total of forty scientists and other staff in its initial phase. Today, Professor Uwe Schneidewind is the president of a team of 240 scientists and supporting staff.

"The Wuppertal Institute undertakes research and develops models, strategies and instruments for transitions to a sustainable development at local, national and international level. Sustainability research at the Wuppertal Institute focuses on the resources, climate and energy related challenges and their relation to economy and society. Special emphasis is put on analysing and stimulating innovations that decouple economic growth and wealth from natural resource use."

This is how the Wuppertal Institute's mission statement describes the Institute's activities. Research focus are the transition processes towards a sustainable development. This requires an integrated approach to policy and science because many of the issues it raises cannot be addressed within a single department or using the tools of individual scientific disciplines. Therefore the four research groups work on:

- Future Energy and Mobility Structures
- Energy, Transport and Climate Policy
- Material Flows and Resource Management
- Sustainable Production and Consumption

As a branch of the Wuppertal Institute, the Berlin Office strengthens the research based consulting that we provide for the world of policy.

An International Advisory Board stands for the independence of research and the Institute's scientific quality. It evaluates the Wuppertal Institute's annual research agenda. The Wuppertal Institute has the legal status of a non-profit limited company (gemeinnützige Gesellschaft mit beschränkter Haftung, according to German law), and is in the responsibility of the Ministry for Innovation, Science and Research of the Land North Rhine-Westphalia. Third-party funding supports most of the Institute's budget and projects.

Address Wuppertal Institut für Klima, Umwelt, Energie GmbH, Döppersberg 19, 42103 Wuppertal, Germany.
Email info@wupperinst.org;
Website http://wupperinst.org/en/home/

University of Freiburg

Mission Statement of the Albert-Ludwigs-Universität Freiburg i.Br

Founded in 1457, the University of Freiburg is one of the oldest German universities and is now one of the nation's leading research and teaching institutions. The University of Freiburg is consciously aware of its intellectual roots in the occidental Christian tradition, especially in the humanism of the Upper Rhine. Building on the original disciplines of theology, law, medicine, and philosophy, it cherishes its mission of passing on the classical cultural heritage to new generations and continuing the southern German liberal tradition. At the same time, the university is dedicated to defining and pioneering new research areas and promoting a strategic interweaving of the natural and social sciences with the humanities.

Freiburg's central position in Europe, in the corner where Switzerland, France, and Germany meet, where it is nestled in one of the oldest cultural landscapes north of the Alps, has had an unmistakable influence on the town. The university is an essential element of the quality of life in Freiburg; both in the academic sphere and in the perception of the general public, the activities of the university are of central importance and highly attractive. From time immemorial, teaching, learning, and research have formed an indivisible whole. A glance at the list of leading academic figures and Nobel prize winners associated with the university, even only of those who taught and researched in the 20th century—featuring well-known names such as A. Weismann, E. Husserl, M. Heidegger, W. Eucken,

H. Friedrich, H. Spemann, H. Staudinger, A. von Hayek, P. Ehrlich, O. Wieland, G. von Hevesy, H. Krebs and G. Köhler—provides evidence enough that the University of Freiburg is and always has been a charismatic center of supreme academic achievement in all disciplines. On this basis, the University of Freiburg views its social task as being situated at the point at which progress, liberty, and responsibility intersect.

The University of Freiburg thus shows consideration for the life situations of women and men in all domains of society and promotes measures for realizing equal opportunity and a higher percentage of female employees, both at the university itself and further afield. In addition to being a general expression of societal responsibility, this policy also serves to further the excellence of research and instruction in all academic disciplines.

Philosophy still forms the center of the great network of the humanities at the University of Freiburg today. The institution owes a great deal of its worldwide recognition to the intellectual excellence of its humanities disciplines.

It is evident and internationally recognized that the University of Freiburg is pursuing a consistent policy to set up cross-disciplinary networks at all levels. In the area of teaching, this involves an above-average proportion of interdisciplinary modules and a consistently cross-disciplinary orientation of the courses at the Master's level. Where research is concerned, this notion manifests itself in the founding of interdisciplinary centers and above-all in the founding of the Faculty of Applied Sciences, which makes the University of Freiburg the only classical university to combine traditional disciplines with the entire spectrum of microsystems technologies.

We strongly believe that top-level research—firmly established on the foundation of a solid grounding in an individual discipline—can only be successful in the interplay between the disciplines. This conviction has led us to persistently stimulate the development of transdisciplinary networks in all of the central research and teaching areas. On the one hand, this strategy will open up new research fields, and on the other hand it ensures that we are optimally equipped for the scientific challenges of the 21st century. The life sciences, which link biochemistry, molecular cell biology, systems biology, applied science, neurosciences, medicine, mathematics, physics, and various humanities disciplines, already constitute a successful and internationally highly acclaimed example of this strategy. This initiative promises outstanding solutions to some of the central scientific problems of our time. In close collaboration—increasingly documented in joint appointments—with external research institutions such as the five Fraunhofer Institutes, the Max Planck Institute of Immunobiology, and the Kiepenheuer Solar Physics Institute as well as partners in the business sector, the university has a major role at the forefront of the development of innovative forms of cooperation and in exploring new research fields.

Notable discoveries and significant technical innovations with genuine practical applications have already resulted from the bundling of these research capacities within and beyond the university. This synergetic effect is reinforced by the university's membership of the European Confederation of Universities on the Upper Rhine (EUCOR), a cross-border tri-national association of the universities of the Upper Rhine, founded in 1989. The University of Freiburg wholeheartedly supports the vision of a single unbounded research landscape in the upper Rhine region in which universities and businesses work together in close collaboration. Collaboration in all areas of teaching and research on the basis of the intensive exchange of students and lecturers, the setting up of joint research programs, reciprocal recognition of study credits and degrees, and the development of bi- or even tri-national university courses are some of the central tasks and objectives of this confederation of universities.

Thanks to these initiatives and preliminary results, and in association with disciplines and faculties which have all been given a first-class rating, the University of Freiburg is confident of maintaining its leading position in the German university landscape. The infrastructural conditions for sustained top-level performance in teaching and research are optimal. The university's leading position in research is simultaneously the basis for an excellent education, which makes the University of Freiburg one of the most attractive academic centers in Europe; its graduates have high recruitment potential as the elite of the academic, economic, and political world of the future.

Address Albert-Ludwigs-Universität Freiburg, Zentrale Universitätsverwaltung / Rektorat, Fahnenbergplatz, D-79085 Freiburg, Germany.

Email info@verwaltung.uni-freiburg.de

Website https://www.uni-freiburg.de/start-en.html?set_language=en

About the Coauthors

Cheryl Desha (Australia) is a member of The Natural Edge Programme (TNEP)
Dr. Matthias Finger (Switzerland), is professor of political sciences at the Swiss Federal Institute of Technology, Lausanne
Karlson 'Charlie' Hargroves (Australia) is founder and leader of The Natural Edge Programme (TNEP) and a member of the Club of Rome.
Dr. Jochen Jesinghaus (Germany), former researcher at the Wuppertal Institute for Climate, Environment and energy, he was a statistician working for Eurostat and is now civil servant, European Ccommission, European Research Institute, Italy
Dr. Irene Kehler (Germany) was a fellow in 1972 of the Heidelberg based group on empirical educational research
Prof. Amory Lovins (USA) is the great pioneer of energy efficiency. He serves as Director of Research at the Rocky Mountain Institute, Snowmass, Colorado.
Dr. Aklilu Lemma (Ethiopia) was Senior Fellow at the UN Centre for Science and Technology for Development, and a member of the Club of Rome.
L. Hunter Lovins (American), weas at the time of writing Factor Four founder and President of Natural Capitalism Solutions, Longmont, Colorado; she is a member of the Club of Rome.
Dr. Michael H. Smith (Australia) is a member of The Natural Edge Programme (TNEP)
Peter Stasinopoulos (Australia) is a member of The Natural Edge Programme (TNEP)
Prof. M.S. Swaminathan (India), is an eminent researcher into agricultural innovation; he served as the Chairman of the United Nations Advisory Committee on Science and Technology for Development.
Dr. Christine von Weizsäcker (Germany), a biologist by training is president, Ecoropa, an environmental NGO, and wife of Ernst Ulrich von Weizsäcker. She coined the term Fehlerfreundlichkeit (error-friendliness).
Dr. Oran Young (USA), is professor emeritus of social sciences at the Bren School for Environmental Science and Management, University of California, Santa Barbara.

About the Author

Ernst Ulrich von Weizsäcker was born on 25 June 1939 in Zürich, Switzerland. In 1965 he obtained a M.A. in Physics from Hamburg University and in 1969 a Ph.D. in Biology from Freiburg University, Germany. In 1972 he became Professor of Biology at Essen University, and in 1975 he became Founding President, University of Kassel. He was director, UN Centre for Science and Technology for Development, New York (1981–1984), Director, Institute for European Environmental Policy, Bonn, London, Paris (1984–19921), and in 1991 he became Founding President, Wuppertal Institute for Climate, Environment, Energy.

In 1998 he was elected as a Member of Parliament, SPD (serving as Chair, consecutively, of the Bundestag Study Commission on Economic Globalization, and of the Environment Committee).

In 2006 he became Dean, Bren School for Environmental Science and Management, UC Santa Barbara, California. 2007–2009 Co-Chair China Council (CCICED), Task Force on Economic Instruments for Energy Efficiency and the Environment. 2009 retired; *pro bono* Co-Chair, since 2007, International Resource Panel, and Co-President, since 2012, of the Club of Rome.

English language publications (with Jochen Jesinghaus): *Ecological Tax Reform*:(London: Zed Books, 1992); *Earth Politics* (London: Zed Books, 1994); (with Amory and Hunter Lovins): *Factor Four. Doubling Wealth—Halving Resource Use* (London: Earthscan, 1997); (with Oran Young and Matthias Finger): *Limits to Privatization* (London, Earthscan, 2005); (with Charlie Hargroves et al.): *Factor Five* (London, Earthscan, 2009). The last three books are also available in Chinese.

Awards 1989: Premio de Natura; 1996: Duke of Edinburgh Gold Medal of WWF Inter-national; 2001: Takeda Award, shared with. F. Schmidt-Bleek; 2008: German Environment Prize; 2010: Federal High Cross of Merit; 2011: Theodor Heuss Prize; 2012: Medal of Merit, Baden Württemberg.

Homepage www.ernst.weizsaecker.de and English website on this book: <http://www.afes-press-books.de/html/SpringerBriefs_PSP.htm>.

Address P.O.Box 1547, 79305 Emmendingen, Germany;

Email ernst@weizsaecker.de

Photo was taken by James Badham in Santa Barbara, California (USA) who transferred the copyright to the author.

About the Book

On the occasion of the 75th birthday of the distinguished environmental scientist, educational reformer, parliamentarian and politically active and critical citizen Ernst Ulrich von Weizsäcker this unique anthology contains previously published thought provoking texts from 1970 to 2013 spanning several scientific disciplines and combining science and political and societal practice. Among them are key parts of three Reports to the Club of Rome he co-authored. Some rather revolutionary contents include a new university curriculum system supporting interdisciplinary achievements; a five-fold increase of resource productivity, allowing to close down nuclear and fossil power plants and political and-economic proposals to make the five-fold increase of resource productivity happen. Von Weizsäcker is a co-chair of the Club of Rome and a co-chair of UNEP's International Resource Panel (IRP). He was president of Kassel University (1975–1980), director of the UN Centre for Science and Technology (1981–1984), in New York, founding director of the Institute for Europan Environment Policy (1984–1991), president of the Wuppertal Institute for Climate, Environment and Energy (1991–2000), member of the Bundestag (1998–2005), chair of its environment institute (2002–2005), dean of the Donald Bren School for Environmental Science and Management, University of California, Santa Barbara, USA (2006–2008).

If you have any concerns about our products,
you can contact us on
ProductSafety@springernature.com

In case Publisher is established outside the EU,
the EU authorized representative is:
**Springer Nature Customer Service Center GmbH
Europaplatz 3, 69115 Heidelberg, Germany**

Printed by Libri Plureos GmbH
in Hamburg, Germany